MORE ON WAR

MARTIN VAN CREVELD

OXFORD
UNIVERSITY PRESS

OXFORD
UNIVERSITY PRESS

Great Clarendon Street, Oxford, OX2 6DP,
United Kingdom

Oxford University Press is a department of the University of Oxford.
It furthers the University's objective of excellence in research, scholarship,
and education by publishing worldwide. Oxford is a registered trade mark of
Oxford University Press in the UK and in certain other countries

© Martin van Creveld 2017

The moral rights of the author have been asserted

First Edition published in 2017
Impression: 1

Published in the United States of America by Oxford University Press
198 Madison Avenue, New York, NY 10016, United States of America

British Library Cataloguing in Publication Data
Data available

Library of Congress Control Number: 2016946642

ISBN 978-0-19-878817-1

Printed in Great Britain by
Clays Ltd, St Ives plc

For those who study war, wherever it is studied, as long as it is studied.

You may not be interested in war, but war may be interested in you.

—Leon Trotsky

ACKNOWLEDGEMENTS

This book has two sets of roots. The first push came back in 2010 when my friend, Major (ret.) Dan Fayutkin, who was then on the Israel Defense Forces' General Staff, asked me to do a short essay on the essence of strategy. I did in fact write the piece; however, the IDF never took up the idea. As a result, the Hebrew version of the piece remains unpublished.

The second push came three years later when the Swedish War College asked me to participate in a conference on military thought they were organizing. I yield to nobody in my admiration for Sun Tzu and Clausewitz. Working on this volume, one of the first things I learnt was how hard writing theory really is. How Clausewitz in particular did it without using a word processor is beyond me. Still I have long felt that, with reason or without, they had not thought through certain things and answered certain questions. Now, I thought, my opportunity had come.

Other people who, in one way or another, have helped me— sometimes without being aware that they did—include Professor Emeritus Eyal Ben Ari; General (ret.) Prof. Yitzhak Ben Israel; Dr. Robert Bunker; Colonel (ret.) Dr. Moshe Ben David; Mr. Niv David; Dr. Steve Canby; Lieutenant Jeff Clement; Professor Emeritus Yair Evron; Dr. Asia Haun; Dr. Yagil Henkin; Mr. Hagai Klein; Major Dr. Lee Ting Ting; Dr. Jonathan Lewy (my beloved stepson); Dr. Edward Luttwak; Mr. Andy Marshall; Mr. Samuel Nelson-Mann; Colonel Prof. John Olsen; Dr. Tim Sweijs; General (ret.) Jake Thackeray; General (ret.) Dr. Erich Vad; Eldad and Uri van Creveld (my beloved sons); and Professor Martin Wagener. Each and all have made a substantial contribution to this book. But for them, it could not have been completed.

When I first undertook this task my dear wife, Dvora, doubted my ability to do it as it should be done. Now that it is complete, I still do not know whether I have convinced her. Never mind, Dvora. But for you this book, like most of the rest, would never have been born. I love you.

CONTENTS

Introduction

The Crisis of Military Theory

1. Studying War

War is the most important thing in the world. When the chips are down, it rules over the existence of every single country, government, and individual. That is why, though it may come but once in a hundred years, it must be prepared for every day. When the bodies lie cold and stiff, and the survivors mourn over them, those in charge have failed in their duty.

So many books have been written on war that, had they been put aboard the *Titanic*, the ship would have sunk without any help from the iceberg. From Thucydides shortly before 400 BC to the present, there have been many excellent military historians. Their contribution to our understanding of war is immense and growing. Yet history and theory are not the same. History focuses on the specific, the transient, and the non-repeatable. It seeks to record what happened, understand why it happened, and, perhaps, where things may be going. Theory, on which more later, tries to discover patterns and uses them to draw generalizations valid for more than one time and one place. It both describes and, at times, prescribes the nature of the subject matter; what its causes and purpose are; into what parts it should be divided; how it relates to all sorts of other things; and how to cope with it and manage it.

In almost every field of human thought and action, good philosophers abound. They have examined their subjects, be they aesthetics

1

or ethics or logic or the existence of God, and dissected them into their component parts. Next they re-assembled them, often in new and surprising ways that helped readers to expand their knowledge and gain understanding. Yet in two and a half millennia there have only been two really important military theoreticians. All the rest, including some who were famous in their own times, have been more or less forgotten.

Names such as Frontinus (c.40–103 AD), Vegetius (first half of the fifth century AD), the Byzantine Emperor Maurice (539–602), Antoine-Henry Jomini (1779–1869), Basil Liddell Hart (1895–1970), and many others matter only to a few specialists in the field. This even applies to Niccolò Machiavelli's *Arte de la guerra* (1521), which for two centuries gave its title to many other volumes in several languages. Yet Machiavelli is remembered almost exclusively because of his political thought rather than for anything he said about war and armies.

The reasons why these and so many other theorists were forgotten are close at hand. War is a practical business—so much so, some have claimed, as to discourage abstract thought about it. It has much in common with playing an instrument or, at the higher levels, conducting an orchestra. Those who wage it do so to gain victory, not to dish up all sorts of insights. In themselves, not even the best theories can save us from the enemy's sharp sword. This fact made most theorists, who were hoping to proffer practical advice to practically minded commanders, focus on how to organize for war, wage war, fight in war, and so on.

As they did so, they often overlooked two facts. First, no two armed conflicts are ever the same. Second, war itself, forming an integral part of human history, is forever changing and will continue to change. The former problem caused many theorists to engage in a futile and/ or pedantic quest for "principles" or "maxims." The latter shackled them to their particular times and places. That again took them into the kind of detail that has long become irrelevant. For example, both Alfred Mahan (1840–1914) and Julian Corbett (1854–1922), the greatest naval theorists of all time, noted the importance of coal in naval warfare. The latter did so just when the Royal Navy was switching to oil, a very different commodity. Countless others have shared the same fate. Seeking to be up to date, they all but guaranteed that they would soon be out of date.

To this rule there have only been two exceptions. The first was the Chinese commander and sage Sun Tzu (c.544–496 BC). The second, the Prussian soldier-philosopher Carl von Clausewitz (1780–1831). Both have had their exits and their entries. At times they were, or were supposed to be, read by everybody with an interest in the subject. At others they were dismissed as too old, too limited, too philosophical, or all of these. Clausewitz in particular has been more often quoted than read and understood. Nevertheless, both are giants who stand out head and shoulders above the rest. In one form or another they will endure as long as war does. If those who claim that war is in terminal decline are right, perhaps longer.

That is not to say that either volume is without problems—especially Clausewitz's *On War*, which, at the time of its author's death in 1831, was mostly a mass of disordered papers. First, neither Sun Tzu nor Clausewitz has anything to say about the causes of war or the purposes for which it is fought. In the case of Sun Tzu, that is because he opens by saying that war is "a matter of vital importance of the state, the province of life or death, the road to survival or ruin. [Therefore] it is mandatory that it be thoroughly studied."[1] From there, while not blaming war on anybody or anything, he proceeds straight to its preparation and conduct.

By contrast, Clausewitz defines war as the continuation of *Politik*, a term that may mean either policy or politics. In Clausewitz's Prussia, discussion of the best way to conduct a war, even submitting formal written objections, was permitted. Yet he never doubted that, once the sovereign or his representative had issued the orders, commanders and soldiers should and would swing into action. Within their own rather narrow domain, so would civilians. At any rate they had to remain calm, as the governor of Berlin, following Prussia's smashing defeat by Napoleon in October 1806, put it. Clausewitz, in other words, ignored both the purpose policy might serve and the question, "why." By so doing, he grossly exaggerated the ability of rulers to start and conduct war for any purpose and in any way they pleased.

Second, neither Sun Tzu nor Clausewitz have much to say about the relationship between economics and war. Sun Tzu at any rate sounds a cautionary note by commenting on the prodigious cost of war. Clausewitz does not even do that. Had he been asked why, no doubt he would have answered that economics, though essential, do not form part of war proper. Strictly speaking, he may have been right. Still, as Friedrich Engels in particular was to point out, so important are economics, "the dismal science," in respect to war, and war in respect to economics, that neglecting them can only be called a grave shortcoming. One could certainly argue that, in World War II and many others, economics did more to determine the outcome than any military moves did.

Third, both writers tend to take the point of view of senior commanders. The examples they use reflect that fact. So does their readership; they do not address every Tom, Dick, and Harry. Sun Tzu's *The Art of War*, like similar Chinese treatises, was never meant for publication. It was kept secret in the archives where only a selected few had access to it. It is probably no accident that the earliest known text was found in a royal grave dating to the second century BC. *On War*, for its part, was initially sold by subscription among Prussian officers.

Proceeding from the top down, both books tend to make war, especially as experienced by the common soldier, appear more rational and more subject to control than it is. They forget that, for every order that is implemented, there are several, perhaps many, which never are. The problem is particularly obvious in Sun Tzu. Like his rough contemporary Confucius he focuses on the elite, treating the rest as mere human material. Saying that, on the battlefield, everything looks like confusion, he omits to add that, to countless combatants of all times and places, it *is* nothing but confusion and the most awful one at that. Often, as long as they themselves are all right, the troops—tired, hungry, exhausted, and afraid—can have little time to reflect on anything beyond this very immediate fact. Never more so then from 1815 on when growing firepower caused drill to be discarded and led to "the empty battlefield." And never more so than in intrastate war between various, often shifting and hard to separate, factions. The possibility that soldiers, let alone civilians, may have their own ideas and that these ideas may influence the conduct of war at all levels is hardly mentioned.

Neither Sun Tzu nor Clausewitz can be accused of having entirely ignored training, organization, discipline, and leadership. However, their discussions are not without problems. The former limits himself to a few aphorisms. The latter's chapters on organization are mostly dated. As, for example, when he writes about the best way to coordinate infantry, cavalry, and artillery. Worse still, in *On War* factors such as "the military virtues," boldness, "perseverance," and so on fall under the rubric of "strategy." That fits neither the author's definition of that term nor ours.

Fourth, both Sun Tzu and Clausewitz come close to ignoring the implements of war, i.e. the field broadly known as military technology. Sun Tzu only has a few words to say about it. Clausewitz does mention it, but only to add that it relates to war as the art of the smith relates to that of fencing. Both authors knew very well that wars were fought with swords, spears, bows, muskets, cannon, and whatever. Both must also have understood that technology—finding the right weapons, as the British military historian and strategist J. F. C. Fuller once put it—is one of the most important factors that determine the shape of warfare at any given time and place. Equally obviously, though, they did not see technology as a fundamental factor deserving profound consideration. The reason, it seems, was that they saw all the armies of their day using roughly the same technology; and that rapid technological progress such as has become almost self-evident from 1815 on did not exist.

Fifth, neither Sun Tzu nor Clausewitz has much to say about staff work, logistics, and intelligence. Some of what Clausewitz does say, e.g. concerning the number of tent-horses an army needs, is badly dated. Yet staff work and logistics are the building-stones of war. To paraphrase the World War II British Field Marshal Sir Archibald Wavell (1883–1950), the combinations of strategy are sufficiently simple for any amateur to grasp. It is by looking after the logistics, defined as the practical art of moving armies and keeping them supplied, that the professional proves himself. Looking at a globe, an armchair strategist may have little difficulty deciding where the nation's carriers should be based. But taking charge of loading a 90,000-ton vessel with the tens of thousands of different items it must take aboard before leaving port requires lots and lots of expertise.

As to intelligence, both authors, each in his own way, only refer to certain aspects of it. Sun Tzu emphasizes the importance of intelligence and explains the various kinds of spies a commander may use to obtain it. However, he has almost nothing to say about its nature or the way it is or should be interpreted. Clausewitz discusses the nature of military intelligence and the role it plays in war. But he barely touches on the way it is obtained. Their discussions of the subject stand in urgent need of being expanded and updated.

Sixth, both Sun Tzu and Clausewitz emphasize *the* most important characteristic of strategy: namely, its mutual, interactive nature and the way that nature determines its conduct. Sun Tzu makes mutuality the lynchpin of his work but only goes so far in elaborating upon this. With Clausewitz, both the nature and the consequences of strategy tend to be lost among a great many other less important topics. That is why a modern discussion of it is needed.

Seventh, neither is interested in war at sea. Perhaps that is because, in their days, neither China nor Prussia were maritime powers. Or else it reflects the fact that, until World War II inclusive, armies and navies were managed by separate offices or ministries. War at sea is probably younger than its land-bound counterpart. Yet it is depicted on 3,000-year-old Chinese and Egyptian reliefs. Starting with the Battle of Salamis in 480 BC, which caused the Persian invasion of Greece to fail, and ending with the great battles in the Pacific in 1944–5, on occasion it has been as decisive as any of its *terra firma* equivalents. But for their command of the sea, the British in 1982 would never have been able to reach, let alone recover, the Falkland Islands.

Eighth, and for obvious reasons, neither Sun Tzu nor Clausewitz addresses air war (including its maritime branch). The same goes for space war and cyberwar. Nor does there seem to be any modern volume that discusses all these fields on an equal basis, linking them with, and placing them within, the more traditional aspects of strategy. On the other hand, and if only because military budgets are going down, at the beginning of the twenty-first century no call is heard more often than the one for "jointness." That is why such a volume is urgently needed.

Ninth, and again for obvious reasons, neither author addresses what, since 1945, has become by far the most important form of "war"—nuclear war. Whether space war, cyberwar (and network war, and culture-centric war, and hybrid war, and effect-based operations, and any number of other kinds that crop up almost daily) are as revolutionary as their originators claim is moot. What ought not to be moot is that nuclear weapons, by casting doubt on the usefulness of war as a political instrument, wrought the greatest revolution military history has ever seen. Like gigantic mushrooms, they cast their shadow over everything else. Nor will they ever stop doing so. Theories which ignore that fact—whether because they were written before the atomic age or through the authors' own fault—do so at their peril.

Tenth, neither has much to say about the law of war. In the case of Sun Tzu that may be because such a thing barely existed, or so some scholars claim. In that of Clausewitz, it is because he dismisses it in a sentence or two. He says that the law in question hardly diminishes the elementary violence of war.[2] Strongly influenced by Prussia's crushing defeat at the hands of Napoleon, the claim is understandable. In some ways it is also correct. Nevertheless, we shall see that law, informal or formal, unwritten or written, plays as great a role in shaping war as it does in shaping any other social phenomenon. Some would say that, since 1945 or so, its importance has been growing. To the point where, in some cases, it has placed strict limits on the war-making abilities of counterinsurgent forces in particular.

Eleventh, neither is much interested in war between asymmetric belligerents. In this context the word "asymmetric" has two different meanings. First, it may mean war between communities, or organizations belonging to different civilizations. In the case of Sun Tzu, this lack of interest rests on the fact that he lived, commanded, and wrote (if he did) during the so-called Period of the Warring States (c.453–221 BC). His career unfolded against the background of constant warfare among very similar polities in what the Chinese used to call "all under heaven" (Ti'an). He may also have been too contemptuous of the "barbarians" to devote a special chapter to them. Clausewitz's focus on intra-civilizational war is indicated by his insistence that European armies were growing alike, making quantity more important than quality. At the time he wrote the military gap between Europe and the rest of the world was increasing day by day; and in any case Prussia was not a colonial power.

But "asymmetric" can also have another meaning. It may refer to a situation where, instead of armies confronting one another, advancing against each other, fighting each other, etc., the belligerents on both sides are of completely different kinds. Irregulars, also known as freedom fighters (partisans), insurgents, rebels, guerillas, bandits, and, last not least, terrorists, may take on armies that are, initially at least, much stronger than themselves. Armies may face irregulars who, initially at least, are much weaker than themselves. Clausewitz at any rate lectured about the problem and gave it a chapter in On War. Sun Tzu did not. Neither provided an intellectual framework capable of containing both symmetric and asymmetric war. Nor does anyone else. The lack of such a framework contributed to, though it did not cause, the inability of the colonial powers to hold on to their empires as well as the American and Soviet defeats in Vietnam, Afghanistan, and Iraq II.

The final reason that made me write the present volume is because, as I can testify from years of experience as a teacher, many young people find both authors hard to understand. With Sun Tzu the problem is the aphoristic style as well as the fact that most of the names mentioned by his various disciples mean nothing to the modern reader. With Clausewitz it reflects the unfinished nature of his work and his sometimes highly abstract way of presenting the material. *More on War* will try to fill the gaps, both those that are self-imposed and those originating in the times and places in which the two men lived and wrote. It will expand on themes which, for one reason or another, they neglected or left untouched, and bring their works up to date wherever doing so seems feasible and worthwhile. All this, in deep admiration and gratitude for what they have accomplished.

2. Practice, History, and Theory

To say it again, waging war and fighting it are practical activities much like playing an instrument or, at the higher levels, conducting an orchestra. Hence one of the best, perhaps *the* best if not the only, ways to familiarize oneself with it is to practice it. As the saying goes, the best teacher of war is war. Other things being equal, the larger and more complex the "orchestra" the greater the role of the conductor, i.e. the commander. It is he who is ultimately responsible for coordinating the efforts of everybody else and directing them towards the objective. All the while taking care that the enemy will not interfere with his plans and demolish them.

Commanders must start by mastering their job at the lowest level. As the ancient historian Plutarch wrote of the Roman general Titus Flamininus (229–174 BC), by serving as a soldier he learnt how to command soldiers. Next, commanders must proceed from one rung to the next until the most competent among them reach the highest level of all. With each additional step additional factors will enter the picture and start playing an increasingly important role. The higher placed a commander, the more varied the kinds of weapons, equipment, and units he will normally be trying to coordinate and use.

Some of the factors are military. Many others are political, economic, social, cultural, and religious. All must be studied, grasped, mastered, and coordinated with all the rest. Such is the nature of war that, at the top, there is hardly any aspect of human behavior, individual and collective, which does not impinge on its conduct. And which, as a result, those in charge do not have to take into account and act upon.

Throughout history some commanders reached their positions and exercised their craft almost purely by virtue of practice combined with their own genius. Take André Massena (1758–1817), the field marshal of whom Napoleon said that he was "the greatest name of my military Empire."[3] Massena was the son of a semi-literate peasant. He served as a sergeant and then spent two years as a smuggler before re-enlisting. He went all the way to the top—without, however, ever entering a military academy. Erwin Rommel, one of the best-known German generals of World War II, did not attend the General Staff College or Kriegsakademie either.

Another example was Ariel Sharon (1928–2014), probably the most gifted operational commander the Israel Defense Force ever produced. Sharon started off as a twenty-year-old private who, in darkness and pouring rain, defended his home village near Tel Aviv against the invading Iraqis in 1948. No sooner did he demonstrate his exceptional abilities than, skipping officer school, he was put in charge of others.

Yet neither experience nor genius suffice. Rarely will the experience of any single person cover all the factors with which, assuming positions of greater and greater responsibility, he must cope. As King Frederick the Great of Prussia (r. 1740–86) is supposed to have said, had experience been enough then the best commander should have been the mule that the Austrian commander Prince Eugene of Savoy (1663–1736) rode on campaign.[4] Worse still, experience can make those who have it impervious to change. The faster things change, the greater the danger.

Geniuses, by definition, are few and far between. Since one cannot predict or command their appearance in history, they are also unreliable. Stalin, at one dark moment in World War II, is supposed to have said, "We do not have a steady supply of Hindenburgs."[5] Whether or not Hindenburg was a genius is not at issue here. What *is* at issue is that, since most of those destined for high command and desirous of exercising it are not geniuses, they will have to do as best they can by relying on study and education instead. As Clausewitz says, from knowledge to capability is a great step. From ignorance to proficiency, a greater one still.[6]

The future is unknown and unknowable. To assume it will be like the past is dangerous—nowhere more so than in war. Nevertheless, study and education can only be based on past experience. Preferably that of others, for in war any lessons learnt must often be paid for with blood. Experience, properly researched, properly organized, and clearly presented is known as history. Of military history Napoleon once said that any aspiring commander should "peruse again and again the campaigns of Alexander, Hannibal, Caesar, Gustavus Adolphus, [the French commander] Turenne, Eugene and Frederick the Great."[7] He should, the emperor added, "model [him]self upon them. This is the only means of becoming a great captain. [His] own genius will be enlightened and improved by this study, and [he] will learn to reject all maxims foreign to the principles of these great commanders."

Note that three of the seven operated over eighteen centuries before Napoleon's own day, in worlds utterly different from his. Many students, accustomed to constant and rapid change and even taking it for granted, will refuse to admit that ancient (and medieval, though Napoleon does not say so) military history can be quite as useful as that which deals with recent events. After all, what can one learn from armies whose most powerful weapon was the pike and whose fastest way of communicating consisted of mounted messengers?

So ubiquitous is this error that not even Clausewitz is exempt from it. The detailed studies he wrote and on which he drew only go back to Gustavus Adolphus in the 1630s. Earlier wars are mentioned in passing and almost exclusively to show how irrelevant they are. Specifically, Clausewitz says, ancient history did not have any lessons to offer "practical" people responsible for preparing war and waging it. Such lessons could only be found in recent wars, which in terms of organization and equipment were more or less like those of his own day. As to what "recent" might mean, he provided not one answer but three different ones. A clearer sign that he never thought out the problem to the end would be hard to find. With all due respect to the master, that is not the way to go.

Furthermore, Clausewitz is wrong in assuming that the main value of military history consists of its ability to provide "lessons" to follow or avoid. Such "lessons" are a dime a dozen. Often they are also contradictory and can be used to prove practically anything. Rather, it is in excavating and bringing to light a reality that, precisely *because* it was very different from the present, can provide us with a basis for comparison. History, especially the history of times long past, is like a magnifying mirror. It can bring out all the blemishes, enabling those who look into it to learn from them and correct them. To use another analogy, the student is like an apprentice who spends some considerable time living and working in a foreign culture. Studying it and immersing himself in it, he may gain a deeper appreciation of the culture in question—that is self-evident—but also, which may be even more important, his own.

Above all, he may realize, must realize, that the familiar is not necessarily all that exists; or can exist; or will exist; or should exist. War is forever changing in a kaleidoscopic, often very rapid, variety of unexpected ways. That is why, if there is anything a commander should guard against, it is the belief, as German Chief of Staff Alfred von Schlieffen (served 1891–1906) once put it, that *das Denkbare ist erreicht* (the conceivable has been attained). Hopefully, studying military history will help protect against that error, and the more remote, the stranger and the less like the present that history is, the better it should so protect. It can, however, do so only if it is studied on its own terms rather than simply as a quest for "origins," "lessons," and "examples."

It is on history that theory is built. Religion ("God's will"), astrology, magic, intuition, and "common sense"—so often used as a substitute for study—aside, it is the only basis on which it can be built at all. To be sure, no theory can possibly do justice to the complexity of life. On the other hand, but for theory any lessons history may offer will remain obscure, causing means to dominate ends. Without theory those, always the great majority, who neither had the opportunity to learn everything by experience nor are geniuses will be unable to distinguish the relevant from the irrelevant. Each time they embark on some enterprise they will have to re-invent the wheel. Even Lord Nelson, Mahan suggested, could have done with a little formal study, especially during his early days when it was not yet available to, or required of, naval officers. Finally, without theory, there is no stock of common ideas, no common terminology. As to what happens when there is no such terminology, see the biblical story of the Tower of Babylon.

Theory should be neither a restatement of the obvious nor a jumble of hard-to-understand—often vague, contradictory, and irrelevant— propositions. Still less should it try to foresee every eventuality that might arise and provide a schoolbook solution. It should be neither a manual nor a list of easy steps towards self-improvement. At its best it is simply an attempt to codify the examples, analogies, and principles that history may offer. It dismantles the subject into its parts; separates the essential from the inessential ones; examines the nature of each; and analyzes their relationship with each other as well as other things. Finally it puts them together again in ways that will enlighten and assist those who peruse it.

A worthwhile theory must meet a number of demands. First, it must be aware of its own limits—the things it can and, above all, cannot do. Second, it must define the subject it deals with as precisely as the nature of that subject allows. Third, it must be as unified, as systematic, as comprehensive, and, yes, as elegant as possible. Fourth, it must be sufficiently detailed to be useful, yet not so detailed as to degenerate into hair-splitting. It must be logical without ever losing touch with the historical reality on which it is based; firm without being dogmatic; and sufficiently flexible to adapt to changing circum- stances as they arise.

All this might make the reader think that the value of Sun Tzu and Clausewitz consists mainly, if not exclusively, in the support they provide for military officers on the way to mastering and exercising their craft. Such a view is understandable but wrong. It is as if we valued Beethoven because the European Union took the final movement of his 9th Symphony as its anthem.

If *The Art of War* and *On War* are as exceptional as they are, then that is precisely because they do so much *more* than merely support commanders by providing them with a compass to guide their actions. They do that, of course, but they also create a sort of map. One which, in spite of the above-mentioned limitations, has proved remarkably able to accommodate change. Their subject matter, war, is the most fearful on earth. Many people would much prefer to contemplate art, beauty, justice, love, and whatnot. Still, the nature of the subject should not prevent these works from being valued at their true worth. Namely, as treasures of the human spirit in the same sense that the works of the above-mentioned philosophers are.

3. The Plan of this Volume

The ultimate objective of this volume is to gain understanding—for myself, perhaps for some others too. But I have also tried to make it useful to some of those who, preparing to bear heavy responsibility, may one day plan and wage war. Not by offering them a list of maxims to follow, but by providing soil which may nourish their own thought. Doing so implied several things. First, I tried to make it more comprehensive than either *The Art of War* or *On War*. Second, assuming that busy people can only devote so much time to study and education, I restricted its length. Hopefully doing so has also made it possible to focus on essentials and avoid going into excessive detail.

Third, I tried to use simple, non-technical, jargon-free language. In particular, I wanted to avoid acronyms. Too often they float about like clots in the bloodstream, threatening to block people's brains. Here, again, I take comfort from the fact that neither Sun

Tzu nor Clausewitz uses them. Simplicity also required using the male form for both sexes (as Strunk and White in *The Elements of Style* recommend, as does William Zinsser in *On Writing Well*). Fourth, I have done my best to combine and balance abstract reasoning, historical examples, and quotations in such a way as to make them shed light upon each other, forming an integrated whole.

To sum up, those are my goals. As to whether they have been achieved—let the reader decide.

I

Why War?

1. Emotions and Drives

S trictly speaking, the causes of war are not part of it any more than an egg is part of the chicken that was hatched out of it. To that extent Sun Tzu and Clausewitz, who each in his own way come close to ignoring the question, cannot be faulted. Still, studying the nature, life, qualities, and behavior of a chicken, it is useful to know that it started life inside an egg and not in the womb of a mammalian female or by cell-division as is the case with bacteria. The same applies to war.

For millennia on end, so intertwined were peace and war that many understood the latter as a given. It was a normal, if fearful, part of human life which did not require further explanation. Most people hated it, suffered under it, and bemoaned its horrors. Homer in the *Iliad* calls it "atrocious, the scourge of man." Ares was "the most hateful" god of all.[1] Nevertheless, mythological tales concerning some long-past Golden Age apart, they resigned themselves to it. In ancient Greece, for example, many peace agreements were not really peace agreements at all. They were, as Thucydides is supposed to have said (but never did), "armistices in perpetual war," explicitly designed to expire after so-and-so many years. As late as 1609, Spain and the Dutch Republic agreed to suspend hostilities for twelve years. Those over, they were supposed to resume and did resume.

So self-evident was war that few could imagine a world without it. That even applied to many "utopian" writers, including Plato in the *Republic* (c.420 BC), Thomas More in *Utopia* (1516), and J. V. Andreae in

Toronto Public Library
Bloor/Gladstone
1101 Bloor St W
416-393-7674

- Checkout Receipt -

Nov 11, 2019 2:27 PM

Christianopolis (1621). One and all, their imaginary communities were specifically, in some cases mainly, organized for waging it. In John Milton's *Paradise Lost* (1667) the angels in heaven exchange artillery fire. It was peace, the ability of considerable numbers of people to live together for considerable periods without resorting to much violence, that had to be explained. This view is still alive and well.

Given this background, it is hardly surprising that some people have believed that the search for causes and motives represented either an academic exercise or pure propaganda. The two were sometimes united in the same person. Take Frederick II of Prussia, the best-known "enlightened" eighteenth-century ruler. At one moment, donning the mantle of the *philosophe* he considered himself to be, he complained about being "doomed to make war just as an ox must plow, a nightingale sing, and a dolphin swim in the sea."[2] How much better to spend time talking to Voltaire at Sans Souci! At others, assuming the role of a power-hungry monarch and commander-in-chief, he gave his cynicism free rein.

The most elementary explanation is that war is the product of man's innate wickedness or, as religious people would put it, sinfulness. "Humans, from the time of youth on, tend towards evil," says the Talmud. "There is deceit and cunning, and from these wars arise," says Confucius.[3] With or without religion in the background, often men let themselves be governed by their worst, most despicable, and diabolical qualities: such as greed, hatred, envy, vengefulness, bloodthirstiness, and cruelty. The outcome is constant and universal distrust and fear. Often, this is all too justified.

Distrust and fear in turn force people to engage in a struggle for more and more power. To quote the great English political scientist Thomas Hobbes (1588–1679), the quest "ceases only in death."[4] Interacting with each other in any number of complicated ways, these and similar emotions are responsible for all that is evil in our world; including war as the worst thing of all.

Turning this logic on its head, many have suggested that war originates in divine retribution. The eighth-century BC prophet Isaiah put it best: for the sins of Judea, he proclaimed, the Lord will "lift up an ensign to the nations from afar, and will hiss unto them from the end of the earth; and behold, they shall come with speed swiftly; none shall be weary nor stumble among them; none shall slumber nor sleep; neither shall the girdle of their loins be loosened, nor the latches of their shoes be broken; whose arrows are sharp, and all their bows bent, their horses' hoofs shall be counted like flint, and their wheels like the whirlwind; their roaring shall be like a lion, they shall roar like young lions; yea, they shall roar, and lay hold of the prey, and shall carry it away safe, and none shall deliver it."[5] The idea remains popular in some places.

Around AD 1900, people often spoke of the "pugnacious instinct." War was the outcome of a drive, originating in nature and planted by her in every man to a greater or lesser degree. It had always made them fight each other and would always continue to do so. Subsequent generations preferred "aggressive" to "pugnacious" and "drive" to "instinct." They also tried to anchor that drive in new biological discoveries: first hormones, then genes, then all kinds of processes in the brain. However, since they too attributed war to an emotion or drive, normally a bad one, that view simply continued the previous ones.

Where it differed was in that, instead of being static, it was linked to the idea of "evolution" and, through it, progress. Discovered by Charles Darwin, evolution came to be applied to human affairs in the form of social Darwinism. Thinkers such as the Prussian general and military historian Friedrich von Bernhardi (1849–1930) argued that war was a method, indeed the supreme method, that nature had devised for selecting "the fittest." People and communities are like sharks. They must swim or drown, grow or decay. In a vulgarized form, this theory was widely accepted—and by no means only in Germany.

Just what "the fittest" meant was not clear. With animals, including chimpanzees as our closest relatives, it might mean a combination of intelligence, physical strength and the health on which it is based, and a tendency, possibly rooted in the individual's hormones or genetic makeup, towards competition, aggression, and dominance.

With humans the situation is more complicated. In our species each person's position on the slippery pole is determined not just by personal qualities but by social factors that antedated birth; in other words, by history.

Descent, inheritance, and wealth often enable those who can claim them to occupy positions they could never have reached by their own efforts alone. Though there are exceptions, often lack of these things will limit what even the most intelligent, the strongest, and the most dominant individual can achieve. Be this as it may, social Darwinists saw life, human life included, as a harsh, pitiless, and ever-lasting struggle. "Nature" required that many beautiful flowers be trodden under foot simply for being small and defenseless. Still it was preferable to the alternative, degeneration and ultimate extinction. Given these alternatives, neither individuals nor communities had a choice.

Linked to the idea of war as the product of, and method by which, evolution proceeds is that of sexual selection. The idea that sexuality, by making females prefer the fittest males of their species over the rest, is critical to evolution is already found in Darwin's own work. Later geneticists expanded on it. War, they argue, does not assist evolution by bringing "the fittest," whatever that may mean, to the top of the social heap. What it does is give the victors greater access to fertile women. They sow their oats and leave their genes behind. In evolutionary terms, that is precisely what success means—"be fruitful and multiply and replenish the earth," as the Old Testament puts it.

Vice versa, as the age-long relationship between Mars and Venus shows, war makes those who fight in it eager for sex. Whether that is because they are separated from their partners, or because of an unconscious wish to leave something behind in case they are killed, or owing to their hormones, is hard to say. An early instance of what the sexual drive can do occurs in the *Iliad*. After ten years in front of Troy the Achaeans had enough. Intent on going home, they defied their leaders' attempts to rally them and rushed back to their ships. Up stood Nestor, a wise old man. Crying out, he told them to stay put until each one had "slept with one of the Trojan men's women."[6] It worked, and they stayed.

Rape has always accompanied armed conflict. At times the victors used it as a weapon to humiliate the losers. But it is not always needed. An eyewitness, the writer Simone de Beauvoir (1908–86), says that no sooner had German troops entered Paris in 1940 than they were surrounded by French women, professional and amateur. Allied soldiers on both sides of the Elbe in 1944–5 also had a ball. British and American soldiers would "buy" women with cigarettes, chocolate, and nylon stockings. The initial orgy of rape over, Soviet ones used bread and salted fish for the same purpose. Many women gained security by surrendering to one, if possible high-ranking, member of a conquering army, who would protect them against all the rest. Voluntary or not, sex spread the DNA of both men and women. The commonest human gene on earth is said to go back to the greatest of all conquerors, Genghis Khan.[7] He who, by one story, once said that the greatest joy in life was to embrace one's defeated enemy's wives and daughters in front of their very eyes.

But cannot war also reflect some of our best and most sublime qualities? War, after all, does not consist solely of some men inflicting death and suffering on others. It does that, and in plentiful measure. But it also demands that men be ready to suffer, sacrifice themselves, and even lose their lives while fighting for a "cause," as English puts it so nicely.

The cause or, if one prefers, myth, may be called God, or king, or country, or flag, or whatever. These things are historically determined and vary from case to case. In the end, their precise nature is secondary. They personify the group—that group whose survival is at stake and for which the individual puts his life in jeopardy. Doing so requires that he do much more than merely perceive and understand the cause in question. He must feel it, experience it, down to the bottom of his existence. He must be overtaken, swept along, carried away as if by a breaker much greater, much better and more worth preserving, than himself. As in "give me liberty, or give me death!"

To some people—all of whom can sleep in their beds only because others stand guard over them—the idea that war flows from what is best and most noble in us is anathema. But that does not mean it is without merit. Perhaps, to the contrary. Some of humanity's greatest philosophers and artists have endorsed it. After all, the fact that the killers put their own lives at risk is precisely what distinguishes war from executions, massacres, and genocides. Confusing these things is a sign of ignorance, unwitting or willful.

Many wars have been accompanied by executions, massacres, and genocides. At times such deeds killed more people than did the conflicts of which they were part. Yet war and atrocities are not the same. Rarely if ever did society honor its executioners. Slinking home, often they were despised and avoided by everybody else. Those who put their lives at risk or died for a cause in which they believed have always been celebrated. And with good reason; for isn't doing so an expression of all that is best and noblest in them?

To Aristotle, who taught Homer (among other things) to Alexander the Great, the highest human quality heroic is excellence (*arête*). It marks the man who, willing to die fighting for a noble cause, leads a glorious life. Much later, thinkers such as Georg Friedrich Hegel (1770–1831) and Friedrich Nietzsche (1844–99) differed on almost everything. Yet they did agree that war is the supreme test both for nations (Hegel) and for individuals (Nietzsche). Both, incidentally, experienced it in person. The former did so when his house was burnt down during Napoleon's 1806 campaign against Prussia. The latter, when driving an ambulance in the 1870–1 Franco-Prussian War.

But for the threat of war, Hegel held, the one force holding communities together would be a combination of selfish "interest" and despotism. But for war, Nietzsche held, there would be scant room for the higher emotions as the soil from which all human culture grows. But for war, mankind would be reduced to an animal life. Scratching the earth, eating, sleeping, having sex, and seeking entertainment. "Bread and circuses," as the Romans put it; "the last man," as Nietzsche wrote. And what is known as human culture would come to a slow, painless, and boring end.

"Who looks on death with unblenching brow/the soldier alone is a free man now" (Friedrich Schiller, 1759–1805, German dramatist and poet). Forgotten are past and future, cause and consequence, punishment and reward. Like mist they vanish, leaving a brilliant clear sky. Freedom, the opportunity to be completely oneself, accounts for the explosion of joy that, in the midst of war's horrors, those who go through them often experience. Not seldom the greater the horror the greater the joy, at least for a time. Some psychologists claim that most people in modern societies, not being accustomed to kill, need to be taught and suffer psychological damage as a result. Yet they also speak of "combat high" and relate it to a flood of adrenalin. Such feelings are by no means new. In the *Iliad* King Agamemnon, "his hands dripping with gore," killed Trojan warriors left and right.[8] All the while "lustily" calling his men to follow him.

Throughout history, for every man who abhorred war there was another who reveled in the "dry-mouthed, fear-purging ecstasy" of it (Ernest Hemingway).[9] Some of them were bad, others good, most average. The medieval knight and troubadour Bertran de Born (*c.*1140–1215) wrote a famous poem about the subject. Many others felt the same. To mention just a few: General Robert E. Lee, President Theodore Roosevelt, the Italian writers Gabriele D'Annunzio and Curzio Malaparte, the famous German writer Ernst Juenger—winner of the Prussian *Pour le Mérite* medal and famous author of *In the Storm of Steel*—and U.S. general George Patton. And this list only includes those who actually went through it. On the eve of World War I Churchill, then Lord of the Admiralty, told his wife how much the gathering storm excited him. Countless others, feeling that their lives were as interesting as bricks and longing to participate in some great epos that would endow them with meaning, agreed.

Glenn Gray (1913–77), the American philosopher and World War II combat veteran, wrote of the "delight in destroying [which]...has an ecstatic character...Men feel empowered by it, seized by it, seized from without."[10] The October 1973 War was probably the most difficult one Israel has ever gone through. More troops were killed per day than in any other. Twenty years later, retired general and subsequent prime minister Ariel Sharon told 120 students (and me) that it had been "great."

All this explains why countless men have always played war games of every sort. Also, why war itself has so often been called the greatest game of all. To be sure, the joy is temporary. As life goes on, its place tends to be taken either by nostalgia—for the comradeship, for the adrenalin—or else by a fierce hatred of war. And of course joy is but one of the many emotions involved. Yet while it lasts it can overwhelm even self-declared pacifists.

The World War I English poet Siegfried Sassoon (1886–1967) described the opening days of the Battle of the Somme as "great fun."[11] To his friend and fellow writer Wilfred Owen (1893–1918) "the act of [going over the top and] slowly walking forward, showing ourselves openly" resulted in "extraordinary exultation." Many warriors of all ages have compared killing with having sex. In every field of human endeavor it is those who love what they do who do it best. Is there any reason why war should be different?

2. From Individuals to Communities

One problem with the above explanations is that war is not an individual activity but a collective one. As such, it can hardly act as an instrument, let alone *the* instrument, for selecting "the fittest" individuals. In any war it is the "fittest" members of society, i.e. young men, who die first. They are killed *because* of their fitness, not in spite of it. For several centuries on end African and Asian societies used to be easily defeated and conquered by the imperialist countries. Later, often with guns in their hands, they cast off the yoke. Surely that doesn't prove they are in any way "fitter" than their former masters. Or perhaps it does?

The role of individual emotions and drives, bad or good, in causing war is equally problematic. Such a thing as a collective psychosis does exist. Nazi newsreels show how a talented demagogue can make hundreds of thousands of people share the same emotions, the most violent ones included, at least for a time. Chinese movies of the Red Guards in action during the "Cultural Revolution" are even more impressive and scary. But raised arms, clenched fists, glazed eyes, snarling faces, parted lips, and mouths that yell slogans such as "Heil Hitler" and "death to the Jews" are one thing. The highly organized, often coolly and deliberately planned and executed, activity known as war is an entirely different one.

True, there is always the danger that war will degenerate into a wild free-for-all. But deadly chaos and war are not the same. Historically, the former may have preceded the latter. Unless iron control is maintained, the latter is always in danger of reverting to the former. That is why many commanders, especially in situations such as victorious sieges which provided plenty of opportunity for murder, rape, and plunder, have often done their best to restrain their troops. Others, particularly Roman and Mongol ones, so disciplined their armies as to turn murder, rape, and plunder themselves into something like fatigue duties, systematically executed under strict supervision. That, indeed, was one reason behind their success.

So great is the gap between individual emotions and communal action that many scholars have tended to focus less on individuals and more on the communities of which they form a part. Some believe it is a question of old men who hate and fear young ones. The latter are bound to win in the end; what better way to get rid of them than by having them slaughter each other? This explanation leaves two problems unsolved. First, though war kills young men, it also pushes other young men to the fore at the expense of their elders. Second, at most times and places, so low was life-expectancy that almost everybody was young. As late as 1450 the sons of English ducal—from *dux*, commander—families could only expect to live to twenty-four.

Most Enlightenment thinkers agreed with Jean-Jacques Rousseau (1712–78) that man's basic nature was neither good nor bad. Such being the case, they blamed war on iniquitous government. Kings and their aristocratic flunkeys treated it as a sort of rite of passage, something they had to engage in if they wanted their peers to take them seriously. As the "War of Jenkins' Ear" (1739–48) between Britain and Spain shows, their stated reasons for doing so could be frivolous indeed.

By some accounts, the real cause of the Seven Years' War (1756–63) was the fact that Fredrick II called Madame de Pompadour, mistress-in-chief to King Louis XV of France (r. 1715–74), by her original name, Madame de Poisson. She hated him, and the rest is history. No wonder war was sometimes called "the game of kings," waged by crowned worthies in disregard, and at the expense, of everybody else.

Two factors favored this kind of frivolity. First, it was an age of consolidation when large territorial states were becoming the norm. Armies operated on or near the frontiers. Mistresses, flunkeys, and kings for the most part remained in their palaces, avoiding any personal risk. King Louis XIV of France (r. 1643–1715), who for over seven decades ruled the largest and most solid territorial state of all, himself drew attention to that fact. The loss of this city or that, he said, would still leave him on his throne. The fact that hundreds of thousands of men had to die, and many more suffer, to satisfy what, in his memoirs, he called his desire for glory does not seem to have bothered him.

Second, rulers and their subordinates of both sexes were protected by class feeling, which created a certain international solidarity among them. It is not in every age that the commander of a besieged fortress will engage in polite banter with his colleague on the other side, telling him that, should he succeed in kissing his (the commander's) mistress anywhere, he would allow him to kiss her everywhere. Or avoid devastating that colleague's private property; or present him with a telescope he had lost; or, on one dubious occasion, invite him to open fire first. Solidarity in turn was buttressed and systematized by contemporary international law as promulgated, above all, by the Swiss lawyer Emmerich Vattel (1714–67).

The most important proponents of the idea that war originated in the internal contradictions within polities were the German philosopher Immanuel Kant (1724–1804) and the American publicist Thomas Paine (1736–1809). Both started with the difference between monarchies and republics. Monarchies were governed by kings and aristocrats who, inheriting their positions, were accountable to no one. Going to war in the ways just described, they used it to exchange provinces, settle inheritances, and avenge insults to what was loosely known as "honor." At times they may have done so simply to amuse themselves. The rulers of republics, being held responsible by their fellow citizens, did not enjoy such license. Those citizens had better things to do than allow themselves to be killed on behalf of their supposed betters' whims. Many subsequent liberals expanded on the theme, claiming that the only way to attain "perpetual peace"—the title of one of Kant's books—was to make all countries democratic.

Many nineteenth- and twentieth-century socialists and communists also saw the roots of war in the structure of the community rather than in the caprices of individuals, however wicked, or sinful, or heroic, or pugnacious, or—to use a term for which English has no substitute—horny they might be. However, Karl Marx (1818–83) taught, communities were not split between willful aristocrats and hapless commoners. They were divided between capitalists and pro-letarians, bourgeois and workers, the exploiters and the exploited. The very nature of capitalism was causing the first group in each pair to grow richer and fewer in number, the second poorer and more numerous.

The capitalists of each country were often short of raw materials. Equally often they could not unload their products on their down-trodden inferiors, who did not have the necessary purchasing power. This forced them to engage in ferocious competition and put them under intense pressure. It was that pressure, rather than the greed of individuals, that accounted for war. That was true both at home, on the frontiers, and overseas, where war took the form of imperialist expansion. Later Lenin and Stalin adopted the same thesis in slightly modified form. But why limit ourselves to the moderns? Twenty-four centuries earlier Plato had written that, beyond a certain point, the contrast between riches and poverty would lead to *stasis*, civil war.

Other social factors can also cause civil wars. One is ethnic, reli-gious, and cultural differences that lead to separatism. The relevant groups may be able to live together in relative peace for decades or more. Next, as Lebanon in 1976 and Yugoslavia in 1991 illustrate, they explode into orgies of hatred and violence. Another is a surplus of young men unable to make a living or find a mate. A regime that is too strong, too arbitrary, and too oppressive may encourage armed uprisings. One insufficiently strong to hold its population in check can also encourage armed uprisings. Probably the most powerful cause, uniting all the rest, is a pervasive feeling that the polity in which people live is unjust. This was the case in Tunisia in December 2010 when the slap a policewoman gave a peddler was heard around the world, starting the "Arab Spring."

These and similar problems can translate into political rivalries which, in turn, can lead to war. But not everybody agrees that war is rooted in the "contradictions," as Marx would have called them, inside polities. Nor, starting with democratic but bellicose Athens, is there much reason to think that some kinds of government are *inherently* more peaceful than others. Instead, from Hobbes through Rousseau to early twenty-first-century "realists," thinkers have pointed to the relationship *between* them. Each time a polity gains in relative power and becomes more secure, its neighbors become less so. And yet, since they refuse to acknowledge an earthly superior, there is no court to which they could turn to settle their differences.

Whether the actors are large or small, few or many, is, in the last instance, immaterial. The greater and more rapid the changes in relative power, and denser the network of interests that binds all the members of the system, the more acute the problem. Is it necessary to add that the same applies to the coalitions which polities sometimes form? War, in other words, is the product of international anarchy. Absent right, might will decide.

3. Purposes, Causes, and Why They Matter

Some of the above explanations are rooted in the nature of man, others in the structure of communities and the relations among them. Still they have one thing in common. They apply to all wars at all times and places, or at any rate to all wars at specific times and places; under absolutism, say, or under capitalism. That is why they cannot explain how *specific* wars came about. One major early twenty-first-century biographer of Adolf Hitler spared no effort trying to show that, almost from birth, he was a bad, very bad, man. So, in his view, were many if not most Germans. That is proved by the fact that, when the Fuehrer came to power, they did not have to be coerced. Allowing themselves to be seduced, they "worked towards" him of their own free will.

The account may or may not be valid. Either way, it does not explain why World War II broke out. Had Hitler and the Germans always been bad? If not, how did they separate themselves from other humans and nations and become so, when and why? Were they less bad in, say, 1933 than they later become? And why, in their badness, did they start this particular war, in this particular way, against these particular enemies and no others? Are not such explanations a mere fig leaf, large enough to cover ignorance but signifying nothing?

Says the Talmud, "grasping at everything, you will attain nothing." To explain the outbreak of specific wars it is not enough to point out underlying causes or factors, whatever they may be. One must show that those factors were not only present but influenced those who made the relevant decisions. Doing so is often, perhaps usually, impossible.

Watching President Kennedy in action during the 1962 Cuban Missile Crisis, American historian Arthur Schlesinger (1917–2007) discovered that many of the factors he had analyzed as part of decision-making processes were simply irrelevant. West German Chancellor Helmut Schmidt (1918–) had a similar experience. In October 1977 a Lufthansa airliner was hijacked, landing in Mogadishu in Somalia. Schmidt responded by sending in the commandos. The mission was accomplished and the passengers freed without loss. Had it failed, it could have led to his fall from power, not to mention the death of many people. Known as *Schnauze*, snout, Schmidt was extremely self-confident and as intellectual a politician as they come. Nobody can accuse him of not having studied and thought about what others had done and what he was doing. Still, as he later wrote, at the moment of decision—to go or not to go—everything he had ever learnt concerning underlying factors vanished like snow under the sun. All that remained was the purpose of the operation, the odds on its succeeding, and what would happen if it did not.

To generalize, war does not just happen. Even if we assume an unintended chain of events, as some claim took place during the weeks leading up to World War I, nothing is "inevitable." At some point somebody has to decide: to halt or to continue. Hitler stopped the invasion of Poland, originally planned to start on August 26, 1939. Surely he could have done the same with the one of September 1?

The considerations behind the decision may be grave or frivolous, correct or erroneous. They may or may not succeed in mobilizing society and cause it to support the war. But they *will* involve specific grievances, specific enemies, specific goals, and, above all, specific cost/benefit calculations. There must be, there will be, an "in order to." Probably *never* did a ruler or commander go to war because man's evil nature pushed him into it; or because of his sublime wish to sacrifice himself and feel free; and so on right down to the internal contradictions of capitalism which had caused international competition to intensify.

Hitler may have been thoroughly bad (though he did not consider himself so). The same applies to the German people. And yet, consciously at any rate, that was not why he started World War II. He did so with a whole range of concrete objectives in mind. He wanted to avenge and reverse Germany's recent defeat; destroy Germany's enemies who, following their failed attempts at appeasement, refused his demands; break the constraints that, during World War I, had almost strangled his people to death; acquire "living space" for them; and secure their future for a thousand years. Causes of any kind had little to do with it. War was a purposeful act, deliberately started to achieve certain objectives. Offensive ones in this case, but defensive in many others.

Stated war aims may not be the true ones. Whether or not they are, such decisions form part of politics. Together, they *are* politics. Are we then to conclude that they alone matter and that "deep" or "fundamental" causes, both those originating in the individual and those buried inside communities or in the nature of the relations among them, do not? That one should give up on them and focus, as far as the evidence permits, on the decisions of those who willed the war and conduct it? There are two reasons why doing so would be a grave mistake; in other words, why causes *matter*. First, no man is an island. Hitler's decision reflected his determination to reach his objectives. Yet his thought, his entire personality, was rooted in the way he and his contemporaries saw things, which, for them, formed "objective" reality.

Particularly important was the idea, which Hitler and many others took for granted, that international life was an unending Hobbesian or Darwinian struggle for resources and power. Also that, as he made clear in *Mein Kampf* as well as his unpublished "second book," it could be decided only by force of arms. Another was the idea that Jews, communists, and Slavs were Germany's sworn enemies and had to be destroyed ere they destroyed Germany. Yet another, that time was running out. Either Germany broke out of its encirclement or else it would go under as a great power. None of these "facts" amounted to decisions. But they did form a master narrative without which any decisions Hitler made would have been inconceivable, indeed meaningless. Briefly, without taking causes into account, understanding both what has happened and what may happen is impossible.

Second, no ruler or commander, however mighty, can do whatever he pleases whenever he pleases in whatever way he pleases for as long as he pleases. The Egyptian Pharaohs, the Roman Emperor Caligula who told his grandmother that he could "do anything to anybody," and Josef Stalin were some of the most absolute despots ever. The last-named one was probably the most powerful man in history. Of him Nikita Khrushchev (1894–1971), the future secretary general of the Communist Party, said that when he said "dance," a wise man danced.[12] Yet not even he could carry along his people from one day to the next. For that, considerable time and effort were required. Nor could he mobilize a large army, enthuse its commanders, motivate its soldiers, and launch it against any enemy he chose at any moment he chose by a pure act of will without reference to anybody or anything.

To do all these things, these and other rulers, both those who were "absolute" and those who were less so, had to start by making sure that their decisions would be in harmony with all, or at least some, of the "deep" forces both of their own time and those applying more generally. Failing to do so, they risked the fate that overtook the Red Army when it invaded Finland in 1939 and the Italian one when Mussolini entered World War II in the next year. As these and countless other cases show, social mobilization is everything. When Sun Tzu, at the very beginning of *The Art of War*, says that in preparing estimates the most important factors are concord between rulers and their people and the favor of heaven, presumably that is what he has in mind.[13]

Imagine a heavy goods train climbing up an incline. It is powered by two locomotives. One, in front, pulls on the cables. The other, at the rear, makes them slacken. So how does the train move? In practice, the load is divided. At any moment some of the carriages are pulled, others pushed. A few in the middle are constantly alternating between being pulled and being pushed. The most cost-effective way to run the train would be for the two locomotives to divide the load in exact proportion to their power. That, however, is an extremely difficult feat. As the behavior of the carriages in the middle shows, perfect coordination is achieved rarely if ever. The longer and the less even the track, the greater the problem. The point may come when the gap between front and rear is simply too large. For lack of coordination, the train comes to a halt.

Other things being equal, the more closely the orders coming from above correspond with the "deep" causes of war, both individual and collective, both perceived and "real," the greater the prospect of success. Success will breed success. Yet a perfect match, like Plato's republic, only exists in heaven. The more time passes, the harder it is to keep it so. Should the mismatch grow beyond a certain point, it will cause the troops to throw their weapons away, the army to fall apart, and the people to refuse their assistance or rebel.

II

Economics and War

1. Sinews and Objectives

Strictly speaking, economics is not part of war. Considered as a separate science, it only emerged at some point towards the end of the eighteenth century. Considered as the basis on which all human life necessarily rests, though, economics goes back to the time when men hunted animals, women gathered berries, and both engaged in exchange. Sun Tzu at one point notes how enormously expensive war is. Clausewitz hardly even does that. Nevertheless, without the appropriate material basis individuals and communities cannot exist. Let alone prepare for war and wage it. Marx's friend Friedrich Engels (1820–95) was perfectly right; nothing illustrates the importance of economics in human life better than war does.

To the simplest societies, going to war presented hardly an economic problem at all. A war leader would announce his decision to raid a neighboring tribe and gather followers. They would take their weapons, most of which they manufactured themselves and most of which were similar to those used for hunting. Forming a raiding party, they took what they needed with them and/or lived off the country. Since nobody demanded or received payment, none of this cost much, if anything. Economic problems could only arise in case the expedition were prolonged, leaving nobody to feed the old men, women, and children by hunting or looking after the family cattle.

With more developed societies the situation was different. Rulers and commanders had to find ways to sustain the troops. Sometimes war could feed war, but that was by no means always the case. Those left behind also had to be fed. As the most active elements in the labor force went a-fighting the standard of living of the remaining population was likely to drop sharply. Yet they could not simply be allowed to starve. Moreover, more sophisticated weapons, especially but not exclusively metal ones, could not be made by just anybody, but demanded sophisticated craftsmanship. And craftsmen had to be paid.

Generally the more advanced and the more monetized the society, and the longer the conflict, the greater the economic demands war made on it. Two factors pushed the process along. First, such societies often used paid troops, either such as were permanently retained or such as were hired each time a war broke out and dismissed when it ended. Second, as technology advanced equipment tended to become more expensive. The advent of the industrial revolution and of mass manufacturing at first interrupted the trend. Relative to the resources at society's disposal the cost of many kinds of equipment went down, thus making possible the vast armies of World Wars I and II. However, after 1945 it resumed.

Pushing in the same direction is the fact that modern armies are not always content to buy off the shelf. They insist that the items they procure meet military standards. Doing so may often be necessary. Whether or not that is the case, there is no question that an Arctic-proof, non-conducting, non-reflective paperclip will be much more expensive than its humble civilian equivalent. Such items account for over half of the Pentagon's procurement budget. All this explains why, throughout history, governments have spent far more on armies and warfare than on anything else. True, in most modern states more money goes for welfare than to the military. But only as long as there are no wars.

If economics forms the indispensable basis of war, how much truth is there in Marx and Engels' claim that all wars can "ultimately" be traced back to economic causes? The answer seems to be that few wars were caused solely by economic problems. But it is equally true that such causes played a role in many, probably most of them.

Relative to more advanced societies, tribal ones tend to be very poor. But poverty does not necessarily discourage war; often, to the contrary. War may be waged for economic objectives such as access to water, hunting grounds, grazing grounds, cattle, and, if the societies in question are settled rather than nomadic, agricultural land. Some, though not all, also used war to increase their human capital, incorporating women and children into their own societies. Women in particular were often valued not only as sexual partners and breeders but for their labor. Things worked, and work, both ways. If one side went to war in order to gain economic assets and resources, the other did the same to retain those they already had.

Tribal societies being poor, they rarely presented more highly developed ones with tempting targets. Whatever the Chinese may have sought beyond the Great Wall and the Romans north of the frontier or *limes*, it was not wealth. Caesar says that beyond the Rhine, which to impress the Germans with Roman power he repeatedly crossed, there was nothing but endless forest. Pressing westward, whites in North America looked for land, not for any products of Indian labor or their accumulated treasure. The reverse was not true. Tribal societies have always envied their settled neighbors whose economies, based on agriculture, yielded a surplus. Those neighbors also had towns where the products of craft and industry were manufactured and stored.

This economic imbalance could lead to war between the two kinds of societies. Poor as the tribal ones were, they often held their own against settled ones, militarily speaking. In the end it was the barbarians who conquered Rome, not the other way around. Only around AD 1350 did the Mongols definitely cease to threaten Europe. Ridding Russia of the Golden Horde took longer still. Warlike tribes often profited from war even without actually engaging in it. They did so by extracting "tribute," which was simply a euphemism for blackmail and extortion.

Settled societies, both rural and urban, also waged war, offensive or defensive, on each other. The most important form of wealth they sought was land, now used for agriculture as well as other purposes. It could be exploited directly, by settlement, or indirectly, through the labor of the conquered. Next were resources such as mines; industrial products; hoards of gold, silver, and other precious commodities; and, once again, human capital. Such societies being much more strongly organized and policed than tribes, they could enslave not just women and children but men too. But for the slaves coming in from all over the Mediterranean, the rise of Rome in particular would have been inconceivable. The final, and ultimately the most important, way of profiting from war was by taxation. As the Roman statesman and orator Marcus Tullius Cicero (106–43 BC) wrote, it was a perpetual fine for defeat.

From his day to the middle of the twentieth century little changed. Rome plundered most of the Mediterranean. The Vikings pillaged northwestern Europe. The Venetians and the Crusaders ransacked Constantinople. The Mongols conquered China and the Mughals, India. The Italian republics fought each other over trade. So did the Spanish and the Portuguese, the English and the Dutch, and, later, the English and the French. The great European colonial venture from 1450 to 1914 was in large part economically driven. As late as 1939–41 the self-proclaimed "have not" Axis countries set out to gain "living space" at their neighbors' expense. Often the quest for riches was successful, at times spectacularly so. Since 1450 alone first Portugal, then Spain, then the Dutch Republic, then Britain, then the U.S. grew rich by war among other things. Almost always the bulk of the profits went to members of the upper classes. To that extent, Marx and Engels had right on their side.

In 1946 the United Nations Charter prohibited the use of force for annexing territory, thus limiting states' ability to gain wealth by such methods. But that does not mean economic objectives no longer mattered. As the 1991 and 2003 campaigns against Iraq showed, some wars continue to be waged at least partly in the hope of gain. If it could be done, as on these occasions it was done, in the name of free trade, so much the better. The role of economic objectives in intrastate wars was equally great. Those in Angola, Chechnya, East Timor, Kurdistan, Mozambique, Nigeria, and the Sudan revolved around oil. Those in Burma, Sierra Leone, and Zaire did so around gems and diamonds; those in Liberia and, once again, Burma around timber; those in Afghanistan, Colombia, and Lebanon around drugs (Hezbollah is said to derive a greater part of its income from drugs than from Iran); and those in Zaire, around several of these commodities combined. Not to mention the sums many militias extract from the populations they claim to "protect." In none of these wars were economic motives the only ones. On the other hand, there is hardly one intrastate war that is *not* motivated, at least in part, by such motives.

2. War and Economic Development

The hope for profiting at the enemy's expense apart, there are four other ways in which war can contribute to economic development. The first is by driving technological progress; the second, by creating economies of scale; the third, by encouraging new and creative ways of raising money; the fourth, by acting as an economic stimulus. Though all four methods are linked, here they will be treated in that order.

Civilian and military technology interact in a number of complex ways. First, the line between the two is often vague, even nonexistent. To quote Khrushchev, a soldier cannot fight with his trousers unbuttoned; hence buttons are strategic commodities. So are clothes, many if not most utensils, vehicles of every kind, communication systems, and countless other things. Second, and partly for this reason, any kind of military technology is necessarily rooted in the civilian one of the surrounding society. Rarely has any country succeeded in developing and operating military technology without an adequate civilian infrastructure. Often the manufacturing methods used in one field can serve in the other as well; as when fourteenth-century bell-founders started producing cannons.

Things could also work the other way around. In the eighteenth century bores developed for hollowing the barrels of cannons were adapted to manufacture cylinders for the first steam engines. The British engineer John Wilkinson (1728–1808) produced both. Much the same applies to the materials—such as steel, first produced on an industrial scale in the middle of the nineteenth century—and techniques used in building warships and the like. Both methods and technical devices were readily transferrable from civilian to military use, if there was any difference at all.

The annals of war bristle with a constant stream of new devices: bows, swords, pikes, shields, armor, catapults, siege engines...all the way to cannons, submarines, aircraft, and the latest unmanned airborne vehicles and robots. As already noted, originally the weapons used in war were the same as those used for hunting. Later the two became almost entirely separate. Changes in the materials of which tools and weapons were made apart, the next most important technological development was the use of non-organic sources of energy such as windmills and water-wheels. They were, however, fixed in geographical space and could not be used in the field. As a result, for some two millennia military technology lagged behind the best available civilian practice.

Around 1890 the pendulum swung in the other direction. No longer did innovation proceed more or less accidentally at the hands of individual inventors, as it had previously done. Instead it was turned into a continuous and often well-organized process. Stimulated by government investment which other fields did not receive to the same extent, military technology started drawing ahead. The list of devices which, rooted in civilian scientific research, were taken over by the armed forces and developed primarily for military use is endless. It includes aircraft, radar, many kinds of electronic equipment, computers, helicopters, ballistic missiles, inertial guidance systems, and nuclear weapons. For almost a century it was commonplace for the most recent technological advances to reach the military several years before they were introduced into the civilian market; as, for example, happened with jet engines.

Some of the military-technological advances generated what was known as "spin-off." Spin-off meant all kinds of new devices and techniques that, though first developed for the military, benefited the civilian world and pulled it forward. But that, too, was not the end of the story. The invention of the microchip around 1980 again changed the relationship between military and civilian technology, causing the latter to overtake the former in many ways. Indeed, armed forces, even when operating in Third World countries, sometimes found themselves in a situation where their irregular opponents used more advanced cellphones, hand-held computers, and so on than they themselves did.

Yet technology is but one aspect of the matter. Armed forces have always made heavy economic demands on the civilian economy that supported them. Here armies and navies must be treated separately. In any given society, the former often formed the largest single organizations in terms of the manpower they employed. They consumed and/or used more food, clothing, and many kinds of equipment than anybody else. As has been pointed out above, the latter, though usually smaller in terms of manpower, were much more capital-intensive. Starting with the ancient warships and ending with today's carriers, the machines they used were among the largest and most complex of all.

Equally important, compared with civilian demand, that of the military tended to be more concentrated. These facts encouraged industrialists to adopt manufacturing methods that exploited economies of scale. Large-scale production reduces overheads, as happened around 1660 when the French Army became one of the first to adopt uniforms. Another important benefit of large-scale demand is the stimulus it provides to the development of interchangeable parts. It is no accident that one of the earliest efforts at standardization can be traced to the fifteenth-century arsenal where Venice built its galleys. Another pioneer in this field was Eli Whitney (1765–1825), the inventor of the cotton gin, who contracted with the U.S. government to produce muskets.

To quote Adam Smith (1723–90), the one thing more important than opulence is defense. Preparation for war, let alone its conduct, often required gigantic sums that could never have been collected in peacetime. Such was the need that it sometimes led to entirely new mechanisms for raising money. A famous example is the establishment of the Bank of England in 1694. The original purpose of the Bank was to help finance the wars against France. It continued to do so throughout the eighteenth century, meanwhile picking up all kinds of other functions too. It was partly because of the Bank that the English economy became the most advanced in the world, a position it retained for almost two centuries.

Finally, war can stimulate not just individual producers but entire economies. A detailed explanation of the way these things work had to wait for the British economist John Maynard Keynes in his famous book, *The General Theory of Employment, Interest and Money* (1936). In essence, his solution to recession is simple: spend, spend, and spend. Even if doing so results in budget deficits. Whether or not they were familiar with his theories, both Hitler and U.S. President Franklin Delano Roosevelt initiated vast rearmament programs. In both their countries it worked, ending the Great Depression, the worst in modern times, as if by magic.

The case of the United States was especially spectacular. Throughout the 1930s the country failed to make full use of its productive facilities and suffered from massive unemployment. Starting in 1940, rearmament generated a boom that went on throughout World War II. It pulled millions back into the labor force and raised living standards. It also laid the foundation for the period of prosperity that followed, perhaps the greatest in all history. Something similar, albeit on an infinitely smaller scale, happened in Israel following the 1967 June War. As defense spending doubled, the depression of the previous eighteen months or so ended practically at a stroke.

Whatever the precise relationship between economics and war, there can be no doubt about the critical role of the former in enabling sophisticated societies to wage the latter. Often, too, war and preparation for it have helped societies to overcome their economic difficulties. This leaves an important question open: is there an upper limit beyond which wealth, far from increasing a polity's ability to wage war, starts undermining it? Starting with Lycurgus and Plato, many legislators, historians, and philosophers, ancient, medieval, and modern, thought so. The way they saw it, the men of poor societies would wage war on their richer neighbors. Having triumphed, they grew rich. Excessive riches caused them to grow soft. Allowing themselves to be governed by women, they lost their fighting edge. Attacked in turn by their poorer but more virile and aggressive neighbors, they ended by collapsing in ignominy. Repeating itself, the cycle formed the stuff of which history was made. Lycurgus' solution was to prohibit the Spartans from using gold and silver. Plato for his part wanted his imaginary state to avoid trade, as far as possible.

Rich societies also tend to have few children. Relative to their size they have fewer men of military age. The remaining ones are unlikely to be good at waging war. Some such societies try to solve the problem by separating thinkers from fighters. The outcome, it is said, is that decisions are made by cowards while the fighting is done by idiots. The more so, as is actually the case in today's "advanced" societies, because they are prohibited from expressing or experiencing the joy of war. Others put their trust in technology, as the mid-fourth-century AD anonymous author of De rebus bellicis (About Things Military) did. Chinese officials, such as Jiao Yu (in the mid-fourteenth century), also seem to have put their faith in technology in this way. To no avail. A century after De rebus was written, the barbarians brought the Roman Empire to an end. Far from defeating the barbarians once and for all, China was conquered by them not once but twice.

To sum up the argument so far, no society more sophisticated than a simple tribal one can wage war without a firm economic foundation. On many an occasion, rulers and polities have taken the offensive in the hope of economic gain—not seldom, with considerable success. Given the right circumstances, war and preparation for it can benefit economies. It does so by pushing technological development, developing economies of scale, improving the mechanisms by which money is raised, and stimulating demand in general. However, there do seem to be *some* limits. Rich societies are often less warlike than poor ones. They may even reach the point where all their wealth avails them nothing. So it has always been, and so, presumably, it will always remain.

3. War and Economic Decline

If war can make a society rich, it also has the potential to impoverish and ruin those who wage it. First, war diverts investment from productive uses to unproductive ones. Instead of swords being beaten into ploughshares, ploughshares are made into swords. Many men and, nowadays, a few women of working age leave the labor force and join the armed forces. There they spend their time preparing for war and fighting it.

Second, so enormous are the economic demands of war that they have often forced those in charge to resort to rather dubious methods of raising money to pay for it. Among them were loans, either voluntary or forced, taxation, inflation, and whatnot. Not all these methods were economically beneficial. Loans ate into current revenue and were not always repaid, bankrupting those who made them. That was how Genoa lost its primacy as a banking center after its largest debtor, Philip II of Spain, suspended payments in 1557, 1575, and 1598. Taxation was often indistinguishable from confiscation. Inflation could wreak as much havoc as any bankruptcy could. At times it brought trade and exchange to a halt, causing immense losses both to individuals and to entire polities.

Summing up these problems, the historian Paul Kennedy (1945–) has spoken of "imperial overstretch." Using the Habsburg and British Empires as his primary examples, he argued that they met their doom because they could no longer pay for the vast military establishment needed to protect their far-flung possessions. The list is easily extended to include the Ottoman, Portuguese, Venetian, and Dutch empires—to say nothing of Rome and Byzantium. But why go back so far? The Soviet Union, trying to keep up with America's much larger economy, spent proportionally far more on defense. Spearheaded by the so-called "Iron Crunchers" of the Soviet military–industrial complex around 1980, its armed forces had become perhaps the most powerful in history. Ten years later the entire gigantic structure collapsed, leaving the country in ruins. Some, Kennedy included, believe the United States is heading in the same direction.

True, the demands of war can help create standardization and economies of scale. However, those demands are highly cyclical. Normally, no sooner does peace return than a period of financial retrenchment sets in. The firms involved in war production are likely to face great difficulties. The choice may well be between keeping them alive by government subsidies or allowing them to go bankrupt. Either way, the outcome is waste and disruption.

War, by instilling a sense of urgency and opening the money spigots, often leads to technological innovation. But new is not always the same as economically useful. A new kind of sword or musket may have enabled its inventors to wage war more effectively. But they did little or nothing to promote the kind of technology on which society's economic welfare depended. The same applied to many, perhaps most, other military-technological innovations. Tanks are, or were, useful fighting machines. So much so that, from about 1940 to 1990, the simplest index of land-power was the number of tanks an army had; and so much so that they turned into the symbol of military strength. Yet their economic uses are, to put it mildly, limited.

Even where spin-off does exist, its economic rationale is often dubious. Would the introduction of computers have been much delayed if World War II had never occurred? How about nuclear energy? Or the Internet? Or the Global Positioning System? Is the best and most economical way to develop a new passenger plane really building bombers first and then adapting them to civilian use? Cannot near-exclusive concentration on civilian technology benefit an economy more than spending vast sums on the military? Judging by the example of Japan from 1945 on, it certainly can.

Other problems abound. Military-technological research, especially during wartime, is often pushed forward at a furious pace. The pressure of time may well compel those in charge to adopt a shotgun approach. Not knowing which project is going to bear fruit, they work on many simultaneously. During the twelve years of the Third Reich no fewer than 1,000 different aircraft were developed to the point where they became flying prototypes. That represents an average of one every four and a half days! Another problem is the wish to accelerate development by embarking on the next stage before its predecessor is completed. Either method can greatly increase cost.

Last not least, war destroys things and kills people. Fields are left fallow, forests cut down, factories ruined, real estate demolished, communications torn up or simply neglected. Many participants and bystanders die of hunger, disease, and enemy action. Others become invalids who have to be supported. War can turn entire provinces, countries even, into deserts with hardly a human being in sight. The Thirty Years' War (1618–48) left about one third of the population of Central Europe dead. World War II, the largest and deadliest ever, killed about 2 percent of the earth's population. People in and out of uniform were burnt, blown to pieces, suffocated, crushed, buried alive, drowned. So vast was the suffering in terms of atrocities endured, crippled and mutilated bodies, and ruined lives that it could not be calculated.

Normally the losers suffer more than the victors. However, the latter do not necessarily emerge scot-free. World War II left Britain victorious but with its economy in ruins. It lost its empire, the largest the world had ever seen. Gradually it turned back into what it had been before 1600; a small, crowded, moderately important island off Europe's shores. The group most affected is likely to be young, fit men, just the kind the economy needs most. Enemy action and every kind of hardship will also kill women, old men, and children, but usually in smaller numbers. As the fate of Pomerania during the Thirty Years'

War and that of Seoul in 1950–3 shows, the greatest sufferers are the regions that witness the most intensive fighting. Especially if they pass from hand to hand, and the more so because outside assistance cannot reach them. Scorched earth policies can increase the damage further still.

At times, such are the losses that victors and losers can hardly be told apart. Recovery may proceed relatively quickly. But there have also been cases when it took much longer to accomplish. Some historians believe that neither southern Italy nor the countryside around Jerusalem ever fully recovered from the effects of the Second Punic War (218–201 BC) and the Great Jewish Revolt (AD 67–73), respectively. The same applies to Baghdad after its sack by the Mongol commander Hugalu in 1258. Early in the twenty-first century, the most destructive wars are the intrastate ones that plague many "developing" countries. Should a full-scale nuclear war, on which more below, ever break out, then quite probably neither victors nor losers will be left with much of a people, or any kind of economy, at all.

III

The Challenge of War

1. What War Is Not

The most important point to note about war is that it is not a Euclidean square or circle. Like every phenomenon that has a history, hence underwent and will continue to undergo change, it is difficult, perhaps impossible, to define. Depending on time, place, and the nature of the society that wages it, now it assumes one form, now another.

More problematic still, over the last few decades the term, and many of those with which it is associated, has expanded in all kinds of unexpected directions. We have grown used to speaking of diplomatic war, economic war, political war, propaganda war, the war on cancer, the war on AIDS, the war on poverty, the war on global warming, and as many other kinds of war as there are letters in the alphabet. Some would add the war between the generations and war between the sexes. Not to mention the "Cold War" that, over four and a half decades, divided the world into two quarreling halves and on several occasions seemed about to bring it to a fiery end. The same applies to many of the terms with which war is associated. Italy's dictator Benito Mussolini (1883–1945) professed to be enamored of war and everything pertaining to it (he did, in fact, spend part of World War I fighting in the trenches). Like Hegel, he saw it as the highest expression of collective life. Like Nietzsche, he also saw it as the highest expression of individual life.

Mussolini never ceased exhorting his fellow citizens to engage in war, excel in it, and sacrifice themselves in it on his orders. *Credere, ubbidire, combattere* (believe, obey, fight), as the official slogan went. Unfortunately for him they did not follow—providing a perfect illustration of the above-mentioned gap between top-level policy and those at the bottom who are supposed to risk their lives in carrying it out. Extending the meaning of the term, *Il Duce* "fought" and, as he claimed, "won" many "battles" such as the "battle for the marshes," the "battle for grain," the "battle for children," and the "battle for steel."

Even this expansion was just the beginning. Many media are almost as enamored of war and military terms as Mussolini was. Hence they keep peppering their products with them as he did his speeches. We have football teams that "battle" their opponents, mountain-climbers who "conquer" summits, archaeologists who go "on campaign," and a great many more. "Strategy" has also expanded. The term goes back to the Greek *stratos*, army, *strategos*, general, *strategia*, the art of the general, *strategema*, ruse, and *Strategicon*, the sixth-century AD handbook that describes all of these. Never used in the medieval West, following the fall of Byzantium in 1453 the term all but disappeared. It was only reintroduced during the last years of the eighteenth century, and then in a slightly different form.

As Jomini, Clausewitz, and others used the term, "strategy" referred to the art of conducting major military operations against a sentient opponent willing and able to fight back. It both described that conduct and tried to lay down rules for it. On the whole, that was where things remained until 1939 or so. Later, though, it started to be applied to non-military struggles too: as in political strategy, economic strategy, industrial strategy, business strategy, and strategy for innovation, to mention but a few. This has now reached the point where the term is often used to describe almost any kind of plan—be it for achieving some national objective; or prospecting for oil; or potty-training a child.

Some of these activities, such as "information war," "psychological war," and "trade war" may accompany war and serve it, undermining the opponent's morale and/or hurting his economy. But since they do not entail the actual use of violence, to call them "war" is misleading. For example, economic competition may be cut-throat. Yet "cut throat" is a metaphor, not a reality. The losers may go bankrupt, even starve. But unless violence is used and competition turns into either crime, on which more in a moment, or war, they do *not* have their throats cut or their property destroyed, nor are they assaulted. By contrast, the mutual use of physical violence, commonly known as "fighting," is the essence of war. As Clausewitz, using one of his best metaphors, says, it is to war what cash payment is to commerce.[1] Towards it everything else tends and leads; on it and its outcome everything else depends; from it everything else ensues.

Other activities do not amount to war because they do not involve a living, thinking opponent capable of fighting back. That includes Mussolini's "battles," whose objective was to increase the production of steel, children, and grain and dry out the Pontine Marshes near Rome. But it also includes the war on cancer, the war on hunger, and many similar things.

The absence of a sentient opponent does not necessarily mean that achieving one's objectives is easy. In fact, judging by the frequency with which it is done, invading some not too powerful neighbor may be easier than solving long-standing economic, environmental, health, and educational problems. At times rulers went to war *because* they could not cope with their countries' problems and in the hope of diverting attention from that fact. A case par excellence was the Austro-Hungarian fear in 1914 that, unless the assassination of Archduke Franz Ferdinand was met by a robust military response, the Empire might disintegrate. What the absence of a sentient opponent does mean is that, natural disasters and accidents apart, the activity, short or long, is entirely under control. That is precisely why such activities, including dangerous ones such as extreme mountain climbing, are not war.

That is not to say that war is the only activity involving violence, strategy, or a combination of both. Some games, notably rugby, American football, and their relatives, are both violent and involve strategy. Others, such as chess, *go*, and many modern board and computer games not only involve strategy but were specifically modeled on war and designed to imitate it as far as possible. That explains why commanders have often used them to simulate war, plan for war, and prepare for war. Among some tribal societies, so much do certain games resemble war as to raise the question whether they are not, in fact, a ritualized form of the latter. Arguably war itself is nothing but the most violent, greatest game of all.

War and games do, however, differ in one crucial respect. War, Clausewitz taught, is, or should be, governed by policy. In fact it *is* a form of policy, one conducted not with diplomatic notes but sword in hand. Its objective is to make the opponent bend to our will, offensive or defensive, wholly so as to put him at our mercy or partly so as to permit a favorable agreement. Games are just the opposite. In them politics, i.e. the process whereby policy is made, is neutralized and put aside. At least temporarily; and at least in the carefully circumscribed place where they are played. In the arena, or tournament field, or area where maneuvers are held, or else on the checkered board, or map, or sand table, individual combatants and teams face one another. Once they do so and the game gets under way they are, or should be, beyond the reach of politics.

True, some war games were expressly designed to accommodate political factors as well as military ones. In other cases games may be used to help policy achieve its goal. When the chief of staff of the German Army, General Ludwig von Beck (served 1935–8) wanted to show Hitler that going to war would be suicidal, he set up a game to simulate a conflict with several neighboring states. The Fuehrer, incidentally, dismissed it as "childish."[2] He fired Beck, pushed forward, and ended up just as Beck had predicted. The Soviets in their heyday, to show their own superiority, were said to have constructed a "chess machine" capable of demolishing all opponents (except Bobby Fisher). Still these games, and countless others like them, neither formed a continuation of policy in the Clausewitzian sense nor were meant to bend anybody's will to that of anybody else.

Games also differ from war in that they have rules. Indeed it is the rules that separate each game from all the rest. The first function of the rules is to define what does and does not count as victory, thus separating games from both politics and war. That is why, as Clausewitz says, in war the outcome is never final. The second is to limit the tools and methods players may use as well as the amount and kind of violence they may employ. Even on the rare occasions when there are no limits, as in the famous Roman gladiatorial games and a few others, still things must be arranged in such a way as to ensure that the spectators are safe. Doing so means that the games, unlike war, may only take place in certain carefully constructed locations or structures. Some of the locations in question, especially those in which trials by battle are held, may be considered sacred. As some episodes in the history of medieval tournaments, and also in that of soccer and American football, prove, where the rules are insufficient, or where they cannot be enforced, there is always a risk that games will escalate into war, become war.

In respect to violence we have gang warfare. In Mexico and elsewhere gangs can have hundreds of members, sometimes more. Fighting each other over such things as territory, resources, and women, they are sufficiently strongly organized to demand, and receive, their members' total commitment. They own and use every weapon including helicopters and home-produced submarines and armored cars. Often they shed as much blood as their means and objectives allow. Nor are they reluctant to take on the government forces sent against them. Still an important difference exists. War is something society, or at any rate large parts of society, through its customs and/or legal system, approves of and engages in. Should it end in victory, then society will celebrate not only the campaign in question but, all too often, war as such. Gang warfare, by contrast, consists of everything society opposes and tries to suppress by every method the legal system allows.

Finally, war is not a contest between individuals. Legal or illegal, serious or not so serious, violent or not, such contests are known as duels. Duels have a long, interesting and, depending on one's point of view, honorable or dishonorable history. Some duels were highly regulated, even ceremonious, others less so. Relative to the number of combatants, some could be as deadly as war itself. Some duels, the best known of which was the one between David and Goliath, took place in front of assembled armies. But neither they nor any other duels *were* war. To the contrary, as the last example shows, in theory at any rate some of them were meant to substitute for it and put an end to it.

Unlike a duel, war is a collective enterprise. First, it is collective in the sense that any harm inflicted on each participant is seen as an injury against the entire group to which he belongs; one for all, all for one. Second, it is collective in that it involves the more or less coordinated efforts of many people. It is true that commanders and warriors often use it for self-promotion. But they are only allowed to do so as long as their antics do not conflict with the goals of the group to which they belong. The fourth-century BC Roman Consul Titus Manlius Torquatus famously had his own son executed for rushing forward contrary to orders.[3] Third, the targets are members of the enemy group or polity. Provided they are, their precise identity is secondary.

The last-named feature is what makes war so tragic. It is the only human activity in which men, nowadays a few women too, who have never met each other kill each other, perhaps for reasons they neither understand nor care about. Briefly, neither the duel nor any of the other activities listed in this section, however much they may resemble war in some ways, *are* war. All they do is borrow some of its terminology and put it to parasitic use.

2. War at the Top

Having considered what war is not, we are now in a position to say what it is. First, in war there must be at least one opponent, or enemy, who fights back. Mutuality implies strategy; that is why a single incapacitating blow is not war. Second, the primary instrument of war is physical violence. That even applies when, to avoid more violence, one side asks for quarter and the other grants it. Third, war is not subject to rules in the same way games are. Fourth, it is, or should be, governed by politics (here understood as (1) the non-juridical process by which power and resources are distributed within communities; and (2) the one by which different communities, refusing to acknowledge an authoritative judge over themselves, pursue their objectives amidst, and against, similar communities). Fifth, the purpose of war is to bend the enemy to our will. Sixth, war and the violence it employs are legal, or at least enjoy the approbation of a considerable part of the society that wages it. Finally, war is not an individual enterprise but a collective one.

The nature of war gives birth to a number of challenges—the greatest and most fearful ones anybody can face. Some challenges are felt most strongly at the top of the hierarchy where military factors interact with a host of others. Others are concentrated mainly near the bottom. Here they will be dealt with, as far as possible, in that order.

The first problem is shaping the link between war and politics. In fact, there is not one problem but three. First, as Clausewitz says, policy should control war.[4] Its task is to prepare war as well as possible, direct operations, and, depending on whether it has been won or lost, either derive the greatest benefit or try to ease the consequences of defeat. Yet doing so is easier said than done. War's very essence, consisting of the mutual infliction of violence, presses towards escalation. One side delivers a blow, the other responds with a more powerful one if he can. One side brings up reinforcements, the other responds in kind. Fear, anger, hatred, vengefulness, and similar emotions raise their ugly heads. They feed each other, form an explosive mixture, and make control all but impossible. Soon, whatever restraints may have existed at the outset, such as not killing prisoners or bombing one another's civilians, are cast off.

Escalation can reach the point where policy, far from taking charge, all but disappears. War, set free, *becomes* policy. The most extreme example is the German–Soviet struggle of 1941–5 when both sides fought each other literally to death. Knowing that there would be no future for the loser, they strained every muscle, used practically every means regardless of law or restraint, and sacrificed their all. Under such circumstances, what was there left for policy to decide? Clausewitz called such a conflict "absolute" war.[5] Having come closer to the absolute than he could ever dream of, we today prefer to speak of "total" war.

Second, policy, though it should direct war, should not demand that the latter do the improper or the impossible. That is because, as Clausewitz says, war has its own "grammar." The grammar's rules consist of the physical capabilities and limitations of men and equipment as well as the maxims of strategy. A ruler who disregards them is like a conductor who, in the name of overall harmony, wants to make drums and trumpets sound like violins. In 1941, so obsessed was Hitler with Leningrad, "the cradle of Bolshevism," as he called it, that he violated the first rule of strategy, i.e. concentration of effort.[6] He tried to capture the city while at the same time marching on Moscow and the Ukraine; only to fail to attain his objectives on any of the three fronts.

A ruler facing elections may try to make his military commanders launch an offensive for which the forces are not yet ready, or else force them to hold onto worthless ground. That is why, other things being equal, the problem may be more serious in democracies than in other regimes less sensitive to public pressure. But it is by no means limited to democracies alone. During the 1870–1 Franco-Prussian War Minister-President Otto von Bismarck and Chief of Staff Helmuth von Moltke repeatedly clashed over the former's right to interfere with military operations or even be informed about them. In the end Bismarck succeeded in keeping control, more or less. Not so his World War I successor, Theobald von Bethmann-Hollweg, who was pushed aside by Hindenburg and his deputy, General Erich Ludendorff.

Third, just where should the dividing line between government and military command run? Obviously having the former dictate the latter's moves in any great detail does not make sense. How a regiment of riflemen, or a flotilla of submarines, or a squadron of aircraft should operate is not a political question but a military one. Equally obviously, there are limits to the independence military commanders may demand or exercise. As General Douglas MacArthur during the Korean War famously found out to his cost.

To complicate things even more, the line is and ought to be flexible. Depending on the nature of the community that wages the war and the kind of war being waged, it will run now here, now there. As a rule, the closer to the beginning or end of any war we get the greater the role of policy and the smaller that of purely military considerations. The first enemy soldier to invade a foreign country, and the last one to leave it, have a great political impact. The thousands who follow their example while hostilities last do so much less, if at all.

Solving the problem demands suitable institutional arrangements as well as an appropriate mindset. Until 1550 or so, the normal method was to unite the posts of ruler and commander-in-chief in a single person. Even later, Frederick the Great and Napoleon did the same. The latter sometimes left Paris for months on end, campaigning in enemy territory hundreds of miles away. As late as 1939–45 Hitler spent much of his time in his various field headquarters. It was there that he received situation reports, held conferences, and issued orders to units as far down the chain of command as corps. Clausewitz himself, no doubt with "the God of War," as he called Napoleon, in mind, favored this solution. But it is not one that modern states often adopt.[7]

However it is done, those at the top will find themselves trying to cope with responsibility. More than any other human activity, war tends to crush the nerves of senior officials and commanders. Problems keep dropping down on them like hail. Some are big, others small. Most are urgent, many unexpected. All come in no particular order and require decisions that often involve life and death. Not having been in command, very few people can imagine what conducting one of those tremendous battles on which depends the fate of nations and countries really means. How did Admiral Jellicoe feel, taking the Grand Fleet to the Battle of Jutland in 1916 knowing he could make the British Empire lose the war in an afternoon?

All this requires an Atlas strong enough to carry the world on his shoulders—and show it. But Atlases can be made to order even less than Hindenburgs. Study and experience can foster and hone the qualities high command requires, but they cannot generate them. One is either born with what the German military in its heyday used to call *Verantwortungsfreudigkeit*, joy in responsibility, or one is not. One either has strong nerves, or one does not. Not everyone can fall asleep before starting a battle, as Napoleon sometimes did. According to one story—perhaps apocryphal, perhaps not—Hindenburg slept not only before and after the Battle of Tannenberg but *during* it too. A similar story is told of General Bernard Montgomery at Alamein in 1942.[8] Some have even suggested that too much academic study, rather than helping a person acquire the necessary qualities, actually makes it harder, sometimes impossible, to do so.

A vast chasm separates peacetime, when bureaucracy reigns and violence is prohibited, from active campaigning when victory becomes the one thing that matters. That is why an officer's record in the former does not always provide a reliable indicator of his performance in the latter. Each time a war breaks out some officers to whom few have hitherto paid attention will suddenly emerge as natural talents and be catapulted upwards. Seizing the initiative, taking the place of fallen commanders and the like, often they catapult themselves. A good example is Scipio Africanus (236–183 BC), who attained high command at the ripe old age of twenty-five. Another is Napoleon's general, the aforementioned Massena, a fornicator and a thief, whose real talent only became evident when war broke out.

The opposite, alas, is also true. The last thing anybody needs is somebody like the German chief of staff, Helmuth von Moltke, Jr. A sensitive personality, he had long had his doubts both about himself and about his supreme commander, Kaiser Wilhelm II. In July 1914 he was also physically ill; two months after the outbreak of World War I he had to resign. In April 1940 the Norwegian chief of staff, General Kristian Laake, learning of the German invasion of his country, suffered a nervous collapse. Both General Yitzhak Rabin, who commanded the Israeli Army in the 1967 Arab–Israeli War, and General (ret.) Moshe Dayan, who served as minister of defense in 1973, were reported, perhaps falsely, to have suffered breakdowns either before the battle or during it.

Many such episodes are very hard, probably impossible, to predict and prevent. Instead, all one can do is select prospective commanders as carefully as possible. Next one must promote them step by step, watch their progress, test them, and entrust them with greater and greater responsibilities. All the while praying that the Peter Principle— which, it is said, causes many a commander to end up one rung higher than his abilities warrant—will not apply.

Next, uncertainty. Uncertainty is very much part of life. It is a condition from which no human has ever escaped, or can escape, or, if he is to avoid being turned into a machine, should escape. As in civilian life, it is due partly to chance, i.e. the impact of factors we do not know and over which we have no control. But the fact that war is waged by two sides, each free to do his utmost to double-cross and surprise the other, increases its role beyond all measure. Not to mention the confusion and the stress that often make those involved fail to see what is and see what is not. Many decisions are based more on intuition than on knowledge. The more so because, even as they are being made, information keeps pouring in. The decision made, things hardly ever do develop as expected, calling for more decisions. Yet had the outcome been certain there would have been no need for war. Thus considered, war is simply a method, a violent method, for settling disputes whose outcome is uncertain.

Some of the methods commanders employ to cope with uncertainty are set out in the section on intelligence. Many people and armies are disconcerted by it; again, given the stakes, no wonder. Doing what they can to avoid uncertainty, they may allow it to paralyze them. But not everybody feels this way. Some, properly organized and trained and enthused, thrive on it and revel in it. Taking it as an opportunity to exercise their initiative and their capacity for improvisation, it makes them perform all the better. Some will even create it deliberately by cutting their communications. General Patton in 1943–4 used what he called the rock soup method.[9] A vagrant found a home for the night and promised his host he would make rock soup. First he boiled water and put a rock in it, i.e. engaged. Next he said that the taste would be much improved if some vegetables, i.e. reinforcements, were added.

Finally, friction. Like uncertainty, friction is an inseparable part of life, impossible to eliminate. Like uncertainty, its impact in war is enormous beyond measure. The sheer size of the forces, human error caused by the immense pressure people are working under, and uncertainty itself all account for this fact. As Clausewitz says, waging war is like trying to walk in water.[10] Every move requires far greater effort, and much more time, than it does on land.

Careful planning, by keeping things simple and building in flexibility and redundancy, can do something to minimize friction. There are also moments when commanders, using all their willpower, must make their forces ignore friction and press forward much as a steamroller crushes the obstacles on its way. At no time is doing so more important than during a pursuit; history knows many instances when commanders, unable to press their exhausted men at the decisive moment, failed to gain the victory that was rightfully theirs. Doing so, however, exhausts the troops. In extreme cases it may bring about ruin.

3. War at the Bottom

One difference between those at the top and those at the bottom is that the latter are far less affected by policy/politics. Except in certain kinds of asymmetric war, on which more later, its impact on the rank and file is normally small, often all but nonexistent. The larger and the more bureaucratically organized an army, the more likely this becomes. In many modern countries soldiers of all ranks are prohibited from participating in politics. At times preaching about the subject may damage their morale. How did Soviet soldiers, stuck in Afghanistan, react to the *politruks'* assurances that they were on an "internationalist" mission?

The burden of responsibility is subjective. Some feel it much more keenly than others. Still, other things equal, as we go down the military hierarchy its weight tends to decrease. To lead a nation is one thing, to lead a squad another. Discipline helps; not the least important of its tasks is to take the weight of responsibility off the shoulders of subordinates and shift it to those of superiors instead. By contrast, uncertainty and friction make their effects felt at every level on which war is waged.

The further down the ladder of command we go, though, the greater the role of what the Germans call the *Strapazen* of war.[11] The term *Strapazen* owes its usefulness to the fact that there is no English equivalent. While varying from time to time and also from one arm of service to another, the word combines a sense of the most intense physical effort and discomfort. To these are added periods of intense boredom; fatigue so great that one can no longer distinguish between sleep and its absence; hunger, thirst, heat, cold, rain, strain, danger, pain, and terror; as well as bereavement, loss, and, not seldom, guilt. All are rolled into one, often to the point where they become indistinct. All have an unfortunate tendency to reinforce each other. As the saying goes, it never rains but it pours.

Everybody knows that, at Agincourt in 1415, the English longbowmen defeated the French knights. But how many also know that the men of both sides arrived on the battlefield half dead with hunger, cold, wet, and fatigue, having just marched 200 miles in twelve days? And that the battlefield itself was a sea of mud? Or take the Wehrmacht in 1941. Most of the troops covered the 600 miles (as the crow flies) from the River Bug to Moscow on foot. Carrying their packs, they slogged through heat, dust, rivers, mud, frost, and snow. All in the face of a tenacious enemy while fighting one ferocious encounter after another. Scant wonder that, coming almost within sight of the Kremlin, they were exhausted. In no other field of human activity are the exertions and deprivations nearly as great. Until late in the nineteenth century, more troops died of them than were killed in battle.

Depending on the period and the size of the army (amongst other factors), some commanders shared everything with their troops. Others, whether because they were keen on their comforts or because they thought control was better exercised from the rear, preferred to stay behind. Never was the contrast between top and bottom more pronounced than in World War I. Senior commanders were ensconced in comfortable country houses miles behind the front. They had orderlies to look after them and dined on the best available food. Some, like Field Marshal Douglas Haig, who commanded the British Expeditionary Force in France from 1915 to 1918, changed their underwear three times a day. For transport they used chauffeur-driven motor cars. For exercise they rode horses. They slept in downy beds and sometimes had their wives or mistresses visit them. The troops, especially front-line ones, might rot away in some muddy hole. So much so that they themselves looked and smelt like ambulant clods. Often they were fed only occasionally, and then with bad-quality grub. Burdened by heavy packs, they carried out most tactical movements marching on foot, often at night so as to escape observation.

In the end, there is no doubt that war is the province of blood, sweat, and tears. How to make the troops cope, not just for a few hours but, if necessary, for weeks, months, and years on end? How to make them march, perhaps repeatedly, into the muzzles of the cannon trained straight at them, ready and more than willing to blast them to pieces? How to keep them going amidst the screams, the infernal noise—greater than that of a thousand devils in hell, says Voltaire—and the confusion, not to mention the vibration and the stench of gunpowder, urine, excrement, and burnt and decaying flesh characteristic of the modern battlefield? Even when, panting and with their hearts racing at 200 beats per minute and more, they hardly any longer know what they are doing? Even when, with adrenalin levels rapidly falling after an action, they are so tired that they can hardly move? Even when the situation appears hopeless and no relief is in sight? Even when their comrades are being killed right and left? How to prevent them from falling apart, both physically and mentally?

To these and similar questions there is one answer: training. Given how important training is, the fact that neither Sun Tzu nor Clausewitz has anything important to say about it is surprising. Training overlaps with study and education, but the two are not the same. Training must come first, education later. Training puts greater emphasis on practice; education, on theory and the history on which it is based. The former takes place in the gym, in the field, in the workshop (technical training), and, increasingly, in simulators and virtual reality. The latter, relying primarily on the oral and written word, is more likely to proceed in class, in the study, and in the library.

As technology and other things change, training will have to keep pace. Not so education which, less tied to specifics, can develop more slowly. Training requires units, weapons, and equipment, which are expensive. For this reason, but also because its results are harder to put on display than numbers of men and equipment, there is often a tendency to skimp on it. By comparison, study and education are cheap. Some would argue that, as a result, they are sometimes overdone. Education should allow theory to be formulated. Theory, along with practice, should guide training, thus closing the circle.

Training should resemble war as closely as possible. As the Jewish historian Josephus Flavius (AD 37–c.100) said of the Romans, "their exercises are bloodless wars, their wars bloody exercises."[12] It must prepare the troops for the horrors of the battlefield, not least its extraordinary brutality. Throwing untrained troops into the fray almost amounts to murdering them. Still there are, or should be, limits. What to call the Japanese use of Chinese prisoners and civilians for bayonet practice? According to Aristotle, the Spartan *agoge*, or training course, was more suitable for beasts than for men. The severed heads, truncated limbs, mutilated and burnt bodies, the sight of the wounded dragging their entrails along, cannot be simulated (except by having medics practice first aid on specially anesthetized pigs, or else in a simulator). Perhaps this is fortunate. So terrible is war that many trainees, had they known what to expect, would have run away as fast as they could.

The best one can do is to use gradualism—much as a football coach will have his team play weak opponents before taking on the big boys. The first level of training any recruit must go through involves discipline, i.e. learning how to obey orders. The second, which is often carried out simultaneously with the first, is the physical one. Physical training, consistently carried out over a period of time, can greatly increase strength and stamina. Up to a point, it can also help teach those who undergo it to endure discomfort and pain. Often the trainees themselves will be surprised by the effect it can have. Next they must learn the use of weapons and other equipment. Especially in modern armies with their complex technology, this may be the most difficult and time-consuming part of all. A soldier who fails to master his tools is seriously handicapped.

After individual training comes unit training, i.e. learning how to work with others as part of a team, a machine almost. Finally, having become proficient, both individuals and units must proceed from one-sided exercises to two-sided ones. They must play all kinds of war games and go on maneuvers. Doing so, they will encounter the kind of opposition that tries to thwart them, double-cross them, fight them, and defeat them at every turn. Encountering opposition, they will learn how to deal with it, outwit it, and overcome it.

At this point those leading the exercise should make sure that friction and uncertainty intrude and make their presence felt. God-like, they can suddenly make a supply column fail to arrive or communications break down. They can introduce a fresh enemy coming from an unexpected direction or change missions in mid-exercise. The objective is to conjure up new and unexpected situations, forcing the trainees to develop initiative and act independently. Each exercise should be recorded as carefully and as completely as the available technical means allow. Each should also be followed by a critique as to what went right and what went wrong—what is commonly known as "lessons learnt."

Provided all this is done as it should be, the outcome will be proficiency. Yet proficiency is not everything. A well-trained unit should develop comradeship—not to be confused with friendship, which is much more personal. It must also gain self-confidence, pride, cohesion, and mutual trust. To the point where, for fear of being disgraced in their comrades' eyes, trainees will risk their lives to support each other and avoid letting each other down. To achieve all this training must be tough, sometimes to the point of physical exhaustion. A whiff of danger is not out of place. As Plato wrote, training that is entirely without risk will turn into a childish game.[13] Finally, lack of practice will cause first fitness and then skills to deteriorate. It must be repeated and the proficiency it creates periodically tested.

Even so, such is the stress of combat that only simple things will work. To counter the stress, some kinds of training must be carried to the point where the troops will act almost like automata without having to waste time thinking. Whatever inhibitions they may have must be overcome. They must be ready to kill without hesitation or compunction so they can kill again if necessary. Training can, and sometimes does, make use of the (not uncommon) joy in killing that certain people experience—"the thrill of the kill," as some have called it. However, so as to avoid creating monsters who are a danger to society and themselves, it must be careful not to turn them into indiscriminate killing machines.

This is a field where World War II, and even more so the Holocaust that Hitler and his henchmen perpetrated in its midst, present a terrible warning. Rudolf Hoess, the commandant of Auschwitz, in his memoirs wrote of the iron-fisted training he and other SS men received.[14] So effective was it that they were willing to do anything to anybody for any purpose, or without any purpose at all. Some were more than willing. If such is the price of victory, then perhaps we should pray to God to show us His mercy by granting us defeat.

IV

Building the Forces

1. Organization

A long with training, perhaps the most important factor leading to success in war is organization. Organization is what turns a mere mass into an effective machine, made of human parts, capable of taking coordinated action towards a common goal. It also establishes the procedures people must follow; thus hopefully minimizing at least the kinds of uncertainty and friction that originate neither in the environment nor in the enemy but within the forces themselves. So well trained were Mamluk horsemen, Napoleon wrote, that each one could take on several French soldiers. But a French unit could defeat a Mamluk host ten times its size; as actually happened at the Battle of Mount Tabor in April 1799.[1]

The organization of armed forces is not, nor can it be, uniform. Depending on their countries' geography, as well as those countries' different policies, economies, enemies, objectives, and so on, armed forces will have to adopt different forms of it. The kind of hostilities in which they engage, e.g. regular warfare, counterinsurgency, and so on, is also important. They will develop different services, different arms, different formations, different command systems, and much more. Some will do their best to remain apart from the rest of society. Others will be so much part of it as to be almost indistinguishable from it.

The earliest "armies" were simply bands made up of all the adult men of a tribe. A formal organization with a clear division of labor barely existed. More sophisticated societies have used more sophisticated methods. Depending on the time and place, some of the men were native-born and/or citizens, others foreigners. Many armies consisted of vassals who performed their feudal duty to their lords, as in the Middle Ages. Others had mercenaries hired for the occasion and discharged after it was over. Some were made up of short-term conscripts supplemented by reservists. Others, including most early twenty-first-century modern ones, consisted of long-term professional volunteers.

Each of these types has had its advocates who favored their own method of recruitment, while denouncing all the rest. In fact, the link between the methods used for recruitment on one hand and military performance on the other has always been fairly tenuous. To make things more complicated, men recruited by different methods have often served side by side in the same armies as, on occasion, they still do. As a rule, the more homogeneous the troops in respect to both origins and method of recruitment, the better.

The first demand made on any military organization is that lines of authority and responsibility be properly distributed within it. One man must command, the rest, knowing their places both in respect to their commanders and to each other, obey. A single bad general is better than two good ones. Perhaps even worse is a situation where, following a victorious war, an army retains a whole galaxy of supposedly excellent generals it cannot get rid of. That was what happened to the French in 1919–40 and to the Israelis in 1967–73; in both cases, generals who had done well in an earlier war did not come up to par when a later one broke out.

Next, a fine balance must be found between centralization and decentralization, leanness and robustness, rigidity and flexibility. An overcentralized force will stifle initiative and let opportunities pass. It may also be unable to respond quickly to an emergency. One that is too decentralized may not be able to function at all. A lean force may not be sufficiently robust to withstand a shock. One that is too robust may be cumbersome to use. Rigidity must not frustrate innovation, and flexibility should not result in loss of focus. To make things more complex still, different circumstances may demand that different mixtures of these qualities be adopted against different enemies at different times.

Other things being equal, the larger and more advanced an army the harder it is to command. The obvious answer is a staff.[2] Traditionally, commanders used their relatives and secretaries for the purpose. The first "Capital staffs," as they were known, were created in the 1760s by such commanders as Frederick the Great and the French Field Marshal Victor-Francois de Broglie. Only later did their prospective members receive formal schooling and become specialized. Initially they were very small. As late as World War II the staff of a U.S. division only numbered some 170 officers and men. By 2010 its size had tripled with no end in sight. Many modern armies groan under their weight. The more so because the specialists, aided by modern technology, like to communicate with each other, skipping the commander and undermining the principle of unity of command. And the more so because the larger and more numerous the staffs, the more they tend to obstruct each other.

Armies tend to be large organizations—very often, the larger the better. As Moses found out when he took the Israelites from Egypt to the land of Canaan, the span of control may easily become too great for the abilities of any single person. Tests show that a good commander can handle up to seven units. However, the stress of battle will cause this figure to go down to three or four. The rest will have to cope as best they can.

The obvious solution is to insert intermediate rungs between the commander-in-chief and the mass of the troops. Unfortunately, though, hierarchies create their own problems. The more elaborate they are, then the more difficult, more time-consuming, and more uncertain the transmission of information becomes, both downwards and upwards. Hierarchies also make it much harder for those at the top to reach and understand those at the bottom.

To cope with this problem, commanders of armies of any size will do well to establish a "directed telescope." By that is meant some kind of mechanism, human, technological, or both, that will enable them to cut through the various rungs and focus on what is going on at the times, and in the places, that interest them most. But directed telescopes should not be used for second-guessing and backbiting. Let alone for disrupting the normal channels of command through which most information must necessarily flow. Should they be misused in such a way, they may do more harm than good.

The simplest armies consist of warriors all of whom carry the same weapons and equipment, be they clubs, or javelins, or bows, or whatever. Being small and homogeneous, such forces are relatively easy to assemble, deploy, and command. As numbers grow and technology develops, specialization will enter the picture. This raises the question of how many troops of each kind there should be, and also how to organize them so as to get the maximum both out of each separately and out of all together. Historically speaking most armies had separate units for each arm, e.g. light and heavy cavalry, swordsmen, pikemen, bowmen, and the like. Often specialists, their commanders included, were drawn from different peoples and nations.

One of the few exceptions to this rule of specialist separation were the Roman legions. Numbering some 6,000 men each, they consisted of heavy infantry, some light infantry, cavalry, and artillery in the form of siege engines (catapults). They also had permanent headquarters. These facts go far to explain their exceptional effectiveness. Nevertheless, as late as the middle of the eighteenth century a prominent commander, the Marshal de Saxe (1696–1750), complained that his contemporaries were reluctant to follow the Roman example.[3] Combined arms units, such as divisions and corps, only started becoming the norm around 1790.

Since then the larger the unit, the more likely it is to consist of a combination of many different arms. It is also much more likely to have a variety of specialists responsible for coordinating with other services. Running parallel to this development, starting in World War II there has been a tendency to push down combined arms to the level of the brigade and even the battalion. That, in turn, is one reason for the proliferation of staffs and headquarters of every kind.

An important question that has always preoccupied armies is what tasks to entrust to the warriors/soldiers themselves and which ones could safely be carried out by others (in general, such as cost much less). Tribal raiding parties were sometimes accompanied by women who acted as porters. Later armies had their camp followers made up of slaves, women, and often children as well. They provided various services such as foraging, supply, cooking, laundry, sewing, medical attention, and sex. Not seldom their number exceeded that of the troops. Often they were put under some kind of semi-military discipline. And always they encumbered the movements of the armies they followed.

The advent of railways after 1830 increasingly forced most of these people to stay at home. The services they had provided were taken over by the military and performed by uniformed personnel. The outcome was that, inside each army, the supporting "tails" underwent vast growth at the expense of the fighting "teeth." The latter went down from 90 percent of the forces in 1860 to about 25 percent a century later. Beginning around 1990, the wheel has been turning in the other direction, with numerous services that used to be provided in-house being privatized. However, growing complexity has meant that the trend towards larger tails has scarcely been affected by this development.

To sum up, a single model of organization to suit all circumstances neither does nor can exist. One may, however, divide present-day fighting forces into two basic kinds with many intermediate ones in between. At one extreme are the regular forces of the most "advanced" countries. They come complete with their commanders-in-chief, staffs, different formations, huge "tails," uniforms, clear separation from the civilian world, and so on. At the other are irregulars, often known as guerrillas or terrorists. In some ways they resemble the tribesmen of old, making do with little or nothing of the above. Common to both is a growing tendency to look to private contractors to provide them with services and even do some of their fighting for them. Looking back on history, from around 1500 to 1945 it was generally armies of the former kind which had the upper hand. Whether this will continue to be the case in the future remains to be seen.

2. Leadership

If organization represents the body of the military machine, then leadership is the soul that fills it with life. Without leadership, any organization, however excellent in other ways, will be utterly useless. If anything, that is even more true in war than in peace. Seen from the top, war allows some men to send others to their deaths—in fact, demands that they do so. Seen from the bottom, the men sent to their deaths are required to obey. This is what makes it unique; no other human activity involves anything of the sort.

Leadership is a democratic term. Its first recorded use in English only goes back to 1821. But that does not mean that its essence is time-bound. Since time immemorial leadership has made use, indeed consisted, of four tools: exhortation, example, rewards, and discipline (including punishments). Each commander must develop his own style by navigating between the four. Doing so he must take into account not only circumstances but the different nature of each subordinate and each unit so as to get the maximum out of them all. It is a bit like playing an organ; one whose buttons, keyboard, and pedals consist of human beings.

Of the four, exhortation is so simple as to be self-explanatory. Provided it is backed by rhetorical and literary skills it can be very effective, at least for a time. Or else commanders of all ages would hardly have bothered to harangue their troops or, later, issue orders of the day; what would the works of Thucydides and Livy be without the speeches they claim to record? Without such skill, it is like shouting down a well—both useless and foolish.

Example, too, is easy to explain. Marching back from India to Babylon by way of the Gedrosian desert in 329 BC, the Macedonian Army all but perished of thirst. Yet when its commander, Alexander the Great, was offered a cup of water he refused it and poured the contents on the ground. This story, and many others like it, speaks for itself. Alexander also gave an example by routinely fighting in person. Once he scaled the walls of an enemy fortress at the head of his troops, got into serious trouble, and had to be rescued. Richard the Lionheart did the same. So did many other commanders from ancient Greece to feudal Japan. In the West, one of the last monarchs to get killed in battle was Gustavus Adolphus of Sweden, who died at Luetzen in 1632.

Later on, the growing size and complexity of armies for the most part rendered that solution impractical. Eisenhower in his memoirs says the troops liked his visits. They saw them, correctly, as a sure sign that any danger was miles away. Nevertheless, to this day the relative number of officers, especially junior officers, killed in action remains an excellent gauge of an army's fighting power. Officers who do not fight, hence do not die, cannot provide an example either.

The third element is rewards. As Napoleon supposedly said, it is with colored ribbons that men are led. The rewards he himself used ranged from an approving tweak of a soldier's ear to the kind of citation, or medal, or promotion, or monetary award, which left the troops gasping. Plato in the *Republic* proposed giving those who excelled in war first right to "love and be loved."[4] There are also collective rewards—flags, standards, campaign streamers, and the like. All have been known since at least Roman times. All are meant to increase cohesion and pride, often with considerable success.

Rewards of every kind will only work if they are justly distributed. The more heroic the deed and the heavier the responsibility those who performed it carry, the greater the rewards that should be attached to it. Or else, giving rise to envy and rancor, they can do more harm than good. As far as possible, bureaucracy should be avoided and a spirit of gratitude and generosity kept up. Finally, promptness is essential. A reward that does not follow the deed reasonably quickly is a reward wasted.

The fourth element is discipline. Napoleon called it the first quality of the soldier, coming before valor. More than its civilian equivalent, military discipline is applied not merely to individuals but to entire units, large and small. That is because holding units responsible for the transgressions of individuals, as long as it is not pushed too far, is a good way to create cohesion. What discipline can do is illustrated by the following story, apocryphal or real. Observing a major exercise, Frederick the Great once asked his suite what the most marvelous thing about it was. The clockwork precision of the various moves, they answered. Wrong, he said. It is the fact that, amidst tens of thousands of heavily armed men, many of whom serve against their will, you and I can stand here in perfect safety.

Discipline falls into two kinds: formal and informal. The former is found in regular forces, the latter in irregular ones—which does not necessarily mean it is less powerful. Here we shall deal with the former, leaving the latter for the chapter on such forces and the kinds of war they wage. Formal discipline is based, can only be based, on written military law. If, as was often the case, the soldiers themselves could not read, then their commanders had to explain it to them. Unlike civil law, military law applies not only within a country's sovereign territory but wherever the forces are. Its purpose is to govern the conduct of the troops, both in peace and in war, so as to ensure that they will do what they are told. Compared to the law of society at large, military law tends to impose more duties on those who are subject to it while at the same time granting them fewer rights. It may also go into details, such as dress, comportment, and so on, which civilian law usually leaves untouched.

The framework having been laid down and understood, the troops must learn to obey those who represent it. Examples are often provided at the spur of the moment and rewards can be distributed relatively quickly. But discipline takes time to instill—more, perhaps, than teaching various military and technological skills. It was not only to improve efficiency but to instill the habit of obedience that drill was invented millennia ago. The armies of China, the ancient Middle East, Greece, and Rome all recognized its importance and took care to have their troops perform it and practice it.

Medieval armies seem to have neglected drill, though the matter is moot. Around 1550 the growing use of infantrymen armed with pikes and muskets caused it to re-emerge. Drill served both disciplinary and tactical ends. Growing, its role peaked during the eighteenth century:

"A battalion of 200 or 250 files," a contemporary wrote, "makes a fine impression as it advances on a broad front...The soldiers' legs, with their elegant gaiters and close-fitting breeches, work back and forth like the warp on a weaver's frame, while the sun is reflected blindingly back from the polished muskets and the whitened leatherwork. In a few minutes the moving wall is upon *you*."[5]

Though no longer relevant to the battlefield, drill retains its importance. It teaches the troops to perform the same task over and over again until they do so promptly—automatically, almost—and precisely down to the smallest detail. The nature of the task is of secondary importance. It may even be completely futile per se, as in strutting to and fro with or without shouting in unison. Since the objective is to make the troops obey orders without thinking, futile drill (when not used to excess) actually has its advantages. Fusing the troops together and making them feel powerful, it may even be enjoyable. As a pre-1914 German ditty put it:[6]

> On a hot summer day/when men are being drilled
> a breeze brings the smell of sweat and leather.
> I like fresh air as much as anyone does;
> But this is even better.
>
> What joy! It is as if one sees
> the army's soul in motion.
> Of this a mere civilian
> can have absolutely no notion.

The best discipline, instilled by these and other methods, is that which is neither needed nor felt—in other words, self-discipline. Besides being the most effective, such discipline requires far fewer resources to enforce. Joining the other elements of leadership, its task is to create the "band of brothers" that Shakespeare's Henry V, and after him Lord Nelson, praised so highly. In reality, as Nelson's own fleet showed, war tends to attract its share, perhaps more, of thugs, adventurers, and adrenalin junkies. If only for that reason, rarely can discipline be entirely separated from punishment or the threat of punishment.

The history of military punishments must be as long as that of war itself. In China at the time of Sun Tzu commanders who retreated without permission were summarily beheaded. So were soldiers who abandoned their commanders. The Romans practiced decimation. When a unit was condemned for cowardice or munity lots were cast. Every tenth man having been selected, he was beaten to death by his comrades. Other armies arrested their troops. They put them on bread and water, lashed them, made them run the gauntlet, and executed them, often in horrible ways *pour encourager les autres*. Most present-day armies do not go that far. However, military punishments maintain their age-old tendency to be more severe than those the law mandates for comparable civilian offenses.

Like rewards, punishments should be just and swift. A punishment on time saves nine. Looking after them, and enforcing discipline in general, is the task of the military courts and police. To counter the extraordinary pressures of war, both are often given powers far greater than those their civilian opposite numbers can wield. That is why, in an emergency, rulers sometimes impose military law on their subjects or citizens. The procedures used in military courts also tend to put less emphasis on the rights of the individual and more on speed. They are simpler and less elaborate; by some measures they are crude.

Though exhortation, example, rewards, discipline, and punishments are all essential for managing an army, all can be overdone. Too much exhortation can become counterproductive. A commander who tries too hard to set an example to his soldiers will end by turning into one, an object of ridicule and contempt. His ability to do his job will also suffer. Too many rewards will result in devaluation, as is said to have happened to the Americans in Vietnam. Discipline is all good and well. However, there is a danger that it will stifle initiative and prevent those subject to it from acting rapidly, decisively, and independently when the need arises. Not for nothing did the German term *Kadavergehorsam*, corpse-like obedience, become notorious.

Punishment, when misused, can also become counterproductive. Severe punishments can cause the victims and their comrades to throw away their weapons and desert. The troops may rise in mutiny, turn around, and kill their commanders. That sometimes happened even in the well-disciplined Roman legions. One must know when to apply punishment and when to turn a blind eye. In one story, Frederick the Great, presented with a deserter, asked him why he had fled. The man answered he was afraid of the coming battle. "Come, come," said the king. "Let us fight another battle today; if I am beaten, we will desert together. Tomorrow." And sent him back to his regiment.[7]

In war, nothing is more important than secrecy. Leadership is an exception to this rule: to be effective, it must operate in the sunlight. Leaders must be actors. As General Patton once told a subordinate, an officer is always on parade. However great the strain, he must always be appropriately groomed and dressed so as to inspire confidence. He cannot afford to be confused, or complain, or hesitate, or put his personal affairs before those of the army he serves. He must show his care for his troops, but not to the point where he cannot function. Or else, sensing weakness, they will despise him and take advantage of him.

Finally, a well-timed word or gesture can work wonders. Alexander the Great's case has already been mentioned. Another is provided by Prussia's chief of staff, Helmuth von Moltke, Sr. In 1866, at a difficult moment in the battle of Koeniggraetz when his country's fate seemed to hang in the balance, he carefully selected the best cigar from a proffered case. Doing so he wordlessly convinced onlookers, including the king and Bismarck, who tells the story in his memoirs, that things were not as bad as they thought and that all was going to be well. The more serious the situation, the more important this kind of showmanship—provided it is not, or at any rate is not perceived as, mere showmanship.

3. Fighting Power

Some of the above factors are more important than others. Historically, the make-up of an army (whether it consisted of natives or foreigners, irregulars or mercenaries, conscripts, reservists, or professionals) seems to have had little impact on its military effectiveness. Machiavelli praised citizen-soldiers, but when those his native Florence had raised were put to the test they refused to fight. Swiss mercenaries had a formidable reputation. But by the same token, the no less formidable Spanish, Swedish, and Prussian armies of the period 1550–1780 recruited their troops from every country in Europe. The French and Spanish Foreign Legions, the British Gurkha Regiments, and, increasingly, the U.S. armed forces are also made up of foreigners.

By contrast, organization, training, and leadership are absolutely essential. An army that has them should develop what Clausewitz calls *kriegerische Tugend*, warlike virtues, and what today is known as fighting power. Fighting power separates the men from the boys. Unlike the initial enthusiasm whose place it takes, it is a dour, steady, plodding sort of thing, all but empty of fear or hope. Perfectly fitting uniforms, polished weapons, smart salutes, ceremonies, parades, briefly the vast array of objects and activities broadly known as the culture of war, personify everything troops and units stand for, what they are. They may also provide observers with some clues as to what morale is like. Yet using peacetime impressions to predict wartime performance is very hard, often impossible. Such impressions can be highly misleading; as Mussolini, who was always trying to gauge his troops' morale by reviewing them and looking into their eyes, found out.

Combat makes cowards out of heroes and heroes out of cowards. Often it surprises not only outsiders but participants too. Brought face to face with themselves, they may or may not "find their legs." The one thing most likely to create steady, reliable fighting power is experience, especially common experience of hardship—and what is harder than battle?—undergone and overcome. Even so armies, like football teams, have their off days, when commanders and troops feel as if they are wearing iron weights. Everything seems to go wrong, and performance falls far below what could reasonably be expected. Especially in prolonged campaigns or wars, such days are certain to occur.

Intense combat and insufficient rest will also cause many men to show signs of psychological stress, ranging from quite mild to very severe. Some ostensibly healthy soldiers will become apathetic, no longer caring whether they live or die. Others will scream in their nightmares, curl up in fetal positions, tremble uncontrollably, go blind, lose the use of their limbs, wet their beds, and become impotent. They see and hear the ghosts of dead comrades. The number affected may exceed that of the wounded. The less well led and cohesive a unit, the more serious the problems. In the very long run no troops and no army are immune against the *Strapazen* of war. Alexander's men spent ten years fighting their way from Macedonia to northern India. Yet even they ended by rejecting his exhortations and demanding that he take them home.

As long as the limits are not overstepped, though, a well-organized, well-trained, well-led and cohesive army, especially when it is supported by a society that knows how to value it, may perform miracles of endurance and heroism far beyond anything conceivable in peacetime. Long distances seem to shrink. Time considered barely sufficient under ordinary circumstances nevertheless sees the mission accomplished. Learning of the rapid French advance towards Berlin in 1806, one Prussian prince is said to have exclaimed, *"Pourtant, ils ne peuvent pas voler"* (but they cannot fly, can they?).[8]

The 1967 Arab–Israeli War provides another excellent example. First Egyptian dictator Gamal Abdel Nasser concentrated his forces on Israel's border and closed the Straits of Tiran. Mobilized in turn, for three weeks Israel's own commanders and troops, most of them reservists, strained at the leash. Thousands roared in chorus, telling Nasser to sit still and wait until they came for him. Some of the roaring was no doubt stage-managed to help conceal and overcome the men's fear, while impressing both the enemy and Israel's own population. Yet as became clear when the signal was given, it rested on a rock-hard core. Storming forward, the Israelis fought as if there were no tomorrow. Many, having watched Arab crowds dancing in the streets and bellowing "death to the Jews," felt certain that, unless they won, for them and everything they held dear there *was* no tomorrow. That was why, to speak with Sun Tzu, heaven itself seemed to favor them.

Repeatedly exceeding their objectives, they kept up the momentum until the end.

A final caveat is in order here. Pent-up fighting power may not actually cause war; for that, a purpose and a decision are needed. But it can certainly generate intense pressure in that direction. On the eve of the 1967 War Israel's generals were eager to break the stalemate that Nasser's aggressive moves had created. So much so that Deputy Chief of Staff Ezer Weizmann burst into Prime Minister Levi Eshkol's room, tore off his epaulets, threw them on the table, and shouted that each day's delay in going to war would result in additional casualties. There have even been extreme cases when such armies were as dangerous to their own side as to the enemy.

Still, in war, having to rein in the noble steeds is usually better than having to prod the reluctant oxen. To speak with Shakespeare's King Henry V:[9]

> In peace there's nothing so becomes a man
> As modest stillness and humility;
> but when the blast of war blows in our ears,
> then imitate the action of the tiger.
> Stiffen the sinews, summon up the blood,
> Disguise fair nature with hard-favored rage.
> Then lend the eye a terrible aspect;
> Let it pry through the portage of the head
> Like the brass cannon; let the brow overwhelm
> as fearfully as doth a galled rock
> overhang and jutty his confounded base,
> Swill'd with the wild and wasteful ocean.
> Now set the teeth and stretch the nostril wide,
> Hold hard the breath and bend up every spirit
> To his full height.

V

The Conduct of War

1. Technology and War

To reiterate, neither Sun Tzu nor Clausewitz were very interested in the relationship between technology and war. Paradoxically, though, their failure to expand on technology, which is forever changing and whose future is all but impossible to predict, may be a major reason why their works have withstood the test of time as well as they did.

By one definition, man is the technology-creating animal. There has probably never been a war fought without weapons and equipment, however primitive. Conversely, even the most primitive weapons and equipment did as much to mold the wars in which they were used as the most modern and sophisticated ones. Swords were no less important in shaping the way Roman legionaries fought than assault rifles are in governing how present-day infantrymen do so. That is why the theoreticians' neglect of the subject is a severe shortcoming. To really appreciate the music an orchestra is producing one has to have a pretty good idea of the instruments it uses, what they can and cannot do, and how they fit in with all the rest. The same applies to war.

The famous German Regulations of 1936 defined war as "a free creative activity resting on scientific foundations."[1] What makes it a free creative activity is the fact that those who wage it are human beings. As such they enjoy considerable freedom to decide and act as they see fit. Material things, in other words technology, are different. Some modern weapon systems, notably those used in air combat, can make certain kinds of decisions without requiring human intervention.

However, the decisions must be pre-programmed. Of "freedom" there can be no question whatsoever.

If war nevertheless rests on scientific foundations, then that is primarily because of the material basis with which, and amidst which, it is waged. A bow has such-and-such a range and penetrating power, but no more. A fighter-bomber can deliver so-and-so many tons of ordnance to such-and-such a distance. A bridge can carry such-and-such a weight. Time, distances, the quantities of supplies needed, and so on can be and must be mathematically calculated. Often he who calculates better wins.

Tools and machines tend to be less flexible and less versatile than their human inventors and operators. No technology can do as many different things, in as many different environments, let alone switch from one activity and one environment to another, as man can. The flying submarine that Benzino Napaloni (Benito Mussolini) in Charlie Chaplin's 1940 film *The Great Dictator* ranted about still has not been built. To compensate for these and other shortcomings, one must have not a single tool or machine but as many different ones as there are purposes, circumstances, developments, and opponents. Not technology as such, but *appropriate* technology. Yet this demand too is not without problems. The more varied the technologies, the more complex their use.

Unlike humans, weapons and equipment have no drives, no feelings, no objectives, no morals, and no morale. They do not require leadership, example, reward, or discipline; as to punishing them, doing so is mere tomfoolery. As long as they are in working order they will perform. They do not refuse orders, they do not forget what they have learnt, and they do not make mistakes. Their attention never wavers and they do not grow tired.

Above all, they can perform many functions much better than humans can, or else they would never have been developed. They have greater power (both offensive and defensive), greater speed, greater reach, greater accuracy, and many other things. Briefly, they act as force multipliers. The advantages they bring are especially important in environments where men without suitable equipment are unable to live, let alone fight: i.e. on the water, under the water, in the air, and in space. Without technology, the newest environment of all—cyberspace—would not even exist.

Nevertheless, hardware on its own is dead material. It will not work, much less reach its maximum potential, without well-trained human operators. At the lower levels that refers to each weapon or weapon system separately. At the higher ones it is a question of integrating and orchestrating them in such a way as to create the best combination. Doctrine and organization must also be adapted. Technology, however excellent, that has been deployed without due attention to all these factors will be all but worthless. In 1870–1, such was the secrecy with which the French surrounded their newly developed mitrailleuse (machine gun) that the troops never learnt how to use it. On occasion, technology may backfire; as by requiring more forces to support and guard it than its performance is worth.

Properly mastered at all levels, fitted into the appropriate doctrinal and organizational framework, and used, technology will vastly enhance its owners' capabilities. Such a situation is sometimes known as technological superiority. At a minimum, technological superiority in the form of better weapons, better transport, and better communications and information-gathering gear should enable its owners to reach further, strike harder, defend themselves better, and operate faster than their opponents. Some forms of technology can even enable those who have it to operate in environments, such as the air or space, where those opponents cannot fight back at all.

Another way technological superiority, properly deployed at the outbreak of hostilities, may contribute to victory is by producing technological surprise. A good example is the October 1973 Arab–Israeli War. Crossing the Suez Canal, the Egyptians used massed anti-tank guided missiles to stop the Israeli counterattack. In the process they wrecked an entire Israeli armored brigade. Whether the surprise represented an intelligence failure or followed from a failure to disseminate and absorb the relevant information is irrelevant. Nine years later in Lebanon the Israelis turned the tables on their enemies. They used their new Airborne Warning and Control Systems to shoot down some one hundred Syrian fighters for the loss of one.

Great or small, technological superiority is most useful in short wars. By definition, when a war becomes protracted such superiority, if it does exist, has failed to bring about a rapid decision. The most likely reason for such a failure is that technology has been used in the wrong manner, under the wrong circumstances, for the wrong purpose, against the wrong enemy. As when heavy, inaccurate weapons such as tanks, artillery, and fighter-bombers are employed against terrorists or guerrillas who offer them no targets. The above-mentioned Israeli campaign in Lebanon, which following the initial success dragged on for no fewer than eighteen years, is an excellent example of just this problem.

The longer the war, the more likely those who find themselves at the shorter end of the stick are to learn and adapt until they no longer are. Normally the easiest way to respond is to develop tactical countermeasures. In 1973 the Israelis, after their initial defeat at the hands of anti-tank missiles, took just a few days to come up with new methods that enabled them to resume the offensive. Another way to respond is to obtain similar weapons, possibly including some captured ones. The first thing Hannibal did after beating the Romans at Trebia in 218 BC was to have his troops adopt captured Roman weapons, which were better than his own.

When two roughly equal parties engage in military-technological competition the outcome is almost certain to be a swing effect. Watching and imitating one another, now one side, now the other, will pull ahead. The development of weapons and armor, siege-engines and fortifications, provide excellent cases in point. During World War I the French, the British, and the Germans all produced a new generation of fighter aircraft about once a year on the average without, however, gaining a decisive technological edge. The outcome was to make quality less important and quantity more so; in the end, the conflict was decided by attrition.

This swing effect has now been active for millennia. Peace, by imposing financial restrictions and posing bureaucratic obstacles, may slow it down. War, by creating a sense of urgency and opening the money spigots, may speed it up. Periods of accelerated change are sometimes known as "revolutions in military affairs." But the quest for superiority is not without problems. First, focusing on the future may cause the present to be neglected. The opposite can also happen; when change is rapid, by the time new weapons reach the battlefield they may already be outdated. Second, change causes disruption and makes it hard to maintain readiness. Third, many weapons have a built-in tendency to become not only much more powerful but much more expensive than their predecessors. The outcome is pressure to cut numbers. Smaller numbers will occasion higher development costs per item, and so on in a vicious cycle.

As the process unfolds the moment may come, repeatedly has come, when weapons become so expensive and so few that one can no longer afford to lose them. Weapons that cannot be lost cannot be used. Monstrous Hellenistic warships; late medieval armored knights, said to be so heavy that they needed cranes to mount their horses; and twentieth-century battleships, hopelessly exposed to submarines and airpower, all provide excellent cases in point. The last-named, designed to command the sea, actually spent most of their time in port where they had to be protected at great expense. In the 1990s most bombers went the way battleships had. Aircraft carriers, the largest and most expensive fighting machines ever built, may represent another example. The continued presence in the arsenals of these and similar weapons is often a sign, not of progress but of conservatism; and, sometimes, decay.

As of the early twenty-first century, many new technologies are just beginning to enter service. Those best known among them are robots. That is not necessarily because they are the most important or because they are going to replace humans on the battlefield anytime soon. Rather, it is the hopes and fears associated with Frankenstein-type monsters that might one day take over, controlling us rather than the other way around. Others are anti-ballistic-missile missiles, non-lethal weapons, laser weapons, stealth weapons, space weapons, and cyber-weapons (also known as malware). Plus the inevitable sensors, data links, and computers that hold everything together and keep becoming more sophisticated and more complex.

Some of the new technologies are operated by the troops in the field. Others are controlled from thousands of miles away by personnel who, being far from any danger, raise the question of whether they are warriors or murderers. Large and small, on land, at sea, in the air, and in space, they are performing an ever-growing number of different missions. From surveillance and reconnaissance to guarding various compounds; and from neutralizing roadside bombs to "wasting" or "taking out" terrorists. They have already led to many changes in organization, training, doctrine, and the conduct of operations, and will no doubt bring about many more still.

Will these and other technologies currently under development cause the art of war to make a fresh start? It is too early to tell. But there are precedents; such as the rapid rise, around 1900, of radio, barbed wire, machine guns, quick-firing recoilless artillery, automobiles, tractors and tanks powered by internal combustion engines, dreadnoughts, submarines, torpedoes, and airpower in the form of both Zeppelins and heavier-than-air machines. They suggest that, confronted with a major innovation, typically the first reaction of the military, and by no means only the military, is to see it as a toy and reject it. Often the commanders in question go out of their way to explain why the technology in question will *never* make a difference.

Next such commanders become interested and try to tack it on to existing forces, as by attaching a few tanks or reconnaissance aircraft to each corps or army. The new technology having proved itself, its pioneers are carried away. "It is going to dominate the world!" they shout. "Everything has changed forever! All previous history has become irrelevant!" Gradually it becomes clear that there is no silver bullet. To realize the full potential of the new technologies they must be extensively tested, trained with, and integrated with everything else so as to maximize their strengths and minimize their weaknesses. Even so, there are always ways to counter them. Finally, as both sides acquire and use the technology, they realize that much has remained more or less as it had always been and that the basic principles still apply.

2. Staff Work and Logistics

To divide an armed force into services, arms, formations, units, and so forth is one thing. To make it work, another. Leadership and discipline are the soul of the force. Staff work, i.e. the military equivalent of administration, fills in the details and translates them into action. A force of any size that gets its staff work wrong will turn into a confused mass and may not even be able to exist. Referring to it, Sun Tzu has a few paragraphs about "calculations." Clausewitz does not have even that. Most other authors have done no better. For every book written about staff work the shelves groan under a thousand dealing with intelligence, strategy, operations, and whatnot. Pen-pushers, it seems, are not very popular.

The commanders of the smallest, simplest tribal armies did what staff work was needed in their heads. The *Iliad* does not mention it. That alone proves the Achaeans could not possibly have numbered 250,000 (or, by another interpretation, 50,000) men as the text, at one point, suggests. Larger, more sophisticated and articulated armies committed their documents to clay, wood, stone, bronze, papyrus, vellum, paper, and other materials. The rules, ordnances, and articles of war had to be drawn up. Ditto organizational structures, fields of authority and responsibility, procedures, and so forth. Men joining the service had to be examined, enrolled, classified, billeted, and paid. Records of transfers, assignments, decorations, promotions, and dis-charges had to be maintained. The same applied to men under arrest, men on trial, wounded men, sick men, and prisoners. Weapons, equipment, and all sorts of supplies had to be procured, stored, registered, and issued. Plans and orders for executing them were written down. The orders were distributed and their reception acknowledged. Reports were submitted, all sorts of correspondence conducted, and diaries—known in Greek as *ephemera*—kept.

Many documents had to be abbreviated, listed, catalogued, and indexed, creating even more staff work. To save time and prevent misunderstandings, acronyms were often used. As much of the material as possible had to be standardized, a task made easier by the invention of print. The value of military archives has always been recognized. In garrison they were kept secret. In camp they were centrally located near the commander's tent. On the march the wagons that carried them were placed in the midst of the column and heavily guarded. During battle they were left behind. Some armies, notably the Roman and Prussian ones, were famous for their paperwork. The Berlin agency responsible for looking after missing Wehrmacht soldiers alone has some 400,000,000 documents of various kinds. With computers dominating the field, one may imagine how large a modern equivalent would have been.

As part of any military operation, looking after the details, which in many cases is just another name for staff work, is absolutely essential. Such work may be conveniently divided into two classes. First, it is necessary to find out what the problems involved are. Second, solutions must be found for those problems. Obviously no staff work that does not do both things can be of any value. That, however, does not mean that staff work should come to an end the moment an operation has begun. First, since few military operations of any size ever proceed like clockwork, problems are certain to present themselves even after D-Day and H-Hour. Second, once an operation is over it will be necessary to look back on it with a critical eye. What went right? What went wrong? Why? Where did it leave us? What next?

Speaking of logistics, Sun Tzu is content with a few generalities. Clausewitz's musings on the subject, while containing a few fine insights, are so focused on the Napoleonic period and so dated as to be useless. Both writers could not be more wrong. As has been said, ungrammatically but accurately, "logistics is the stuff that if you do not have enough of, the war will not be won as soon as." That is as true at the higher levels as at the lower ones. The Prussian Army in which Clausewitz served illustrates this truth. Prussia started as the smallest and weakest of the great powers. The outcome, a built-in tendency to neglect logistics in favor of operations, was carried over into the German Army. In both World Wars it made a major contribution to defeat.

Logistics are the art of the possible. It is logistics that ultimately determine the maximum size of an army; whether it can operate; in what season it can operate; where it can operate; the distance it can get away from its bases, and for how long; the speed at which it can move; and so on. "This is logistics," as one officer put it, not magic or gung-ho wishful thinking. But one should not go too far. In the end logistics should be the handmaiden of operations, not the other way around. An army that waits for the last button to be sewed onto the shirt of the last soldier will never get going. There are also situations in which commanders, seizing an opportunity and taking a risk, should press forward almost regardless of what their quartermasters say.

Logistics have accompanied war from the beginning. The earliest "armies" consisted of raiding parties numbering, perhaps, a few dozen warriors. Water apart, their main requirement was for food at about three pounds per man per day. That figure has still not changed very much. Warriors lived off the country they crossed while also using any booty they could take. On occasion entire tribes did the same. For example, during the great migrations of the early Middle Ages; or when the Manchurian tribes conquered China in 1644. The larger an army and the more concentrated its mode of operations, the harder it is to feed. The point could easily come where a district or province could no longer provide sufficient supplies, forcing the army to move or starve.

Living off the land, especially over time, also creates other difficulties. Compensating those made to provide food and other supplies is expensive. However, if one does not pay then those affected will surely conceal whatever they can and resist as much as they can.[2] Such a policy, all but forcing the population into the enemy's arms, can be counterproductive. Pillage is the most time-consuming, least efficient form that any logistical operation can take. Many, perhaps most, supplies will be destroyed or wasted. In 1941–3, so inefficient was German management in the east that even tiny Denmark provided the Reich with more food than the occupied Russian lands did.

Even that is only the beginning. A commander who allows his troops to go where they please and take what they need is likely to lose control over them. The outcome will be bad discipline, desertion, and disintegration. Napoleon's *Grande Armée*, which relied on requisitioning, did not deteriorate that far. But it was notorious for the number of marauders who, having left their units, were always following it. Not only were these soldiers missing from the order of battle, but they made relations with the local population much worse than they should have been. If living off the land is unavoidable, the best method is to have the supplies gathered by special parties under tight command. Where payment cannot be offered one can promise it by issuing receipts.

The standard way to deal with these difficulties is to rely on magazines or bases. Even raiding parties sometimes used them, pre-depositing supplies along the route they intended to take. Eighteenth-century armies established magazines to "spring a surprise," as the saying went, on the enemy by starting operations earlier in the season than would have been possible without them. Other things equal, the larger the army the more it depended on magazines. Yet magazines too have their problems. Some products will spoil over time. Some kind of transport system is needed to link the magazines with the forces in the field, restricting the latter's operational freedom. The transport is also certain to be very expensive. As Sun Tzu noted, the greater the distance the higher the cost.[3]

The simplest transport systems relied on the backs of men (sometimes, women). In some kinds of terrain that continues to be the case. Next came animals. Either they carried supplies or hauled them in carts. But animals eat much more than humans do. Feeding them from base has normally been possible, if at all, only over fairly short distances and for short periods; even more than food, fodder had to be gathered from the country. This again limited the size of armies as well as the times and places at which they could operate. As late as the mid-eighteenth century it was thought that the optimal size of an army was 40,000–50,000 men.[4]

Bread apart, an army needs many kinds of equipment. Some of that equipment can likewise be gathered from the surrounding areas. However, the less developed the country, and the more specialized the equipment, the less likely it is to be found. Different kinds of terrain, such as mountains, tundra, forests, and swamps, also make a difference both in themselves and because they tend to be sparsely populated. Today, as in 1798 when Napoleon crossed the Sinai Peninsula on his way from Egypt to Palestine, deserts represent the most difficult terrain of all. In them there is nothing to be had, not even water. During the American War in Afghanistan, water took up more space, and weighed more, than any other kind of supply.

Until World War I, food and fodder used to account for the bulk of military supplies, far exceeding everything else. Modern technological warfare has changed the situation. It has vastly increased the amount of ammunition, fuel, lubricants, and spare parts for the numerous machines operating on the modern battlefield. By some calculations, the daily weight of supplies needed to keep a soldier in the field has grown thirtyfold. Some kinds of equipment, such as weapons and tools, last a long time and need only be replaced if they are lost to enemy action or accidents. Others are consumed or expended and need to be constantly replenished. All this creates problems which only careful planning and strict traffic discipline can solve. In the Gulf War in 1990–1, so busy was the highway carrying supplies to the Americans' left wing that it could be crossed only by helicopter.

Even greater than the growth in the amount of supplies was the increase in the variety of items of which they consist. They range from food to ammunition and from fuel to every kind of communications gear, spare parts, medical supplies, and so on. In the Gulf in 1991, the U.S. forces alone drew from an inventory of five million different items to stay operational and carry out their mission. Many items had different lifespans. They had to be stored under different conditions, looked after in different ways, and distributed by different forms of transport to different units, some of which were moving about at high speed. As the war proceeded many of the units were scattered all over the theater of war. Such diversity, in turn, occasions formidable administrative problems which only the most up-to-date computers and data links can handle.

The above may make the reader think that wartime military administration and logistics are broadly similar to their peacetime equivalents. In respect to many of the technicalities, that view is correct. The more so because, before the rise of cyberwar, interfering with the opponent's administration was all but impossible. But the situation in respect to logistics is in fact different. The weather and unforeseen events such as natural disasters apart, peacetime logistics can proceed without fear of interruption. Not so military systems which must always take the enemy into account. This causes the emphasis to change from efficiency and profitability to defense and survivability.

The impact of these problems on military logistics is dramatic. Foraging parties, depots, lines of communication, and convoys must be carefully placed, dispersed, concealed, and/or protected against attack. The system must be made sufficiently flexible that it can follow the units as they suddenly switch from one mission and one objective to the next. Flexibility in turn requires a considerable amount of slack. The "just on time" systems that businessmen favor can be useless, even dangerous in this context, with the final outcome one of numerous complications, constraints, and extra costs.

Furthermore, the logistical requirements of armies have always presented their enemies with opportunities. Foraging necessarily led to dispersion. It enabled the detachments responsible for it, occasionally entire armies, to be taken by surprise. Districts lying in the enemy's way could be stripped bare, as was done in in the Palatinate in 1689–97, Bavaria in 1704, and Russia in 1707, 1812, and 1941–2. Bases located too close to the front could be captured or destroyed. The same applied to trains. The American Civil War in particular is famous for the way both sides mounted deep raids behind the front, intercepting convoys and cutting railway lines. After 1914 such raids were replaced by air attack. At times, as in Normandy in 1944 and during the Battle of the Bulge during the same year, with devastating effect.

Everything else being equal, the larger the logistic "tail" the more vulnerable it is. So enormous are the logistic demands of modern armies that, by creating new vulnerabilities, they may even have helped encourage new kinds of forces waging new forms of "asymmetric" war. On those forms of war, more below.

3. Intelligence in War

Everything in war is murky. The murk is made even murkier by the fact that both sides, to conceal their moves while misleading and deceiving the enemy, do their utmost to make it so. Just three things can penetrate the murkiness ("uncertainty," as Clausewitz calls it) and disperse it, if only a little: intelligence, intelligence, and intelligence. Espionage in all its endless forms is said to be the second oldest profession. Without it armed forces would be deaf and blind. That explains why, in 2014, the U.S. was spending an estimated $53 billion on foreign intelligence alone.[5] That is more than some 95 percent of what 194 countries on earth spend on all their defense needs combined.

The purpose of intelligence is twofold. We need it to prepare our own plans, as in selecting targets, ranking them in order of priority, and deciding how to reach them. But we also need it to discern and forestall the enemy's moves. Either way, there is no substitute for intelligence. As Sun Tzu points out, neither divine omens, nor crystal balls, nor historical analogies can tell us where our opponent is.[6] Nor in what condition he is, how strong he is at any given time, and how we can best hit him and kill him. Above all, they cannot foretell what the future will bring: what the enemy's intentions are; what he is going to do next; and when and where and how he is going to hit us. The one way to gather information on such matters is by observing the opponent and entering his councils.

Saying that information about the enemy's future plans is absolutely essential does not mean it is the only kind we need and that any other is worthless. Plans, those of the enemy included, are never made, cannot be made, in a vacuum. Willingly or not, they are based on, and reflect, that enemy's thoughts, beliefs, capabilities, and general situation. Without a good understanding of that situation, those capabilities and those beliefs and those thoughts, much of the information we may be able to obtain about his intentions will be incomprehensible.

All this makes gathering intelligence resemble putting together an enormously large and complicated puzzle. A few pieces are colorful and striking, many others drab and uninteresting. Some are so general that they are of little use for answering specific questions. Others, to the contrary, are so specific as to be almost meaningless. Only by putting all of them together, each one exactly in its due place, can a more-or-less complete picture be created. Even then it will almost certainly not be valid for very long.

The earliest tools used in intelligence gathering were humans. Scouts could be posted. Patrols could be dispatched, spies embedded, messengers intercepted, passers-by and prisoners interrogated, and so on. Such methods go back to the Old Testament and further. More advanced civilizations added writing. Written messages were in some ways easier to intercept than oral ones. That in turn led to the development of the earliest codes. Since then the race between code-makers and code-breakers has continued without interruption.

Around the middle of the nineteenth century the first electronic communications were introduced. They enabled information to be passed inconceivably faster than the fastest armed forces could move. The first electronic messages went by wire. No sooner did they start doing so than the wires started being tapped. Intercepting wireless messages, which spread in all directions, was much easier still. Wireless could also be used for direction finding and traffic analysis—finding out, as far as possible, who communicated with whom, how often, and so forth. During World War I all the most advanced belligerents used all these methods, as well as their opposite, radio silence, as a matter of course. They do so still.

Along with telegraphy came photography. It freed commanders from their dependence on their scouts' artistic talents—previously they were often taught to draw. Since photographs are much easier to make than drawings, it also enabled them to obtain and use far more visual information than before. After 1900, the role of photography was much augmented by the use of aircraft. They enormously extended the areas over which intelligence could be gathered and the speed at which it could be gathered. Soon aerial photography, along with the eyes of pilots, became an important source of information. The same applied at sea. During World War I sonar, a device for submarine-hunting, was introduced. Later it was developed to the point where it formed a global network.

World War II saw the introduction of radar. Since then any number of other systems has been added. Some, such as satellites and drones, are well known even though their exact capabilities are often kept secret. Others remain hush-hush. All are capable of translating whatever information they gather into electronic signals. Those in turn can be instantly transmitted to any point it is needed. The entire vast system is connected by the usual inconceivably complex network of computers and data links. The links and computers themselves then become involved in intelligence-gathering, an activity known as "information war."

Not all pieces of information have been born equal. The first factor that governs their value is completeness. Incomplete information that only includes some of the relevant factors may easily be misleading, even dangerous, causing the recipient to take the wrong action or none at all. In practice the information we obtain is *never* "complete" even in peace, let alone in the murky business known as war. Where pieces of the puzzle are missing, as they almost always will be, it is necessary to rely on informed guesswork.

Second, timeliness. War is a dynamic activity; a day, an hour, may make a big difference. In air combat, which is conducted at very high speed, even fractions of a second may do so. This problem may be divided into two parts. First, it is essential that the information, at the time it is gathered, should be up to date. Second, transmission, evaluation, and dissemination must proceed as rapidly as possible. Modern technological means, many of which transmit information at the speed of light and enable it to be received and acted upon in "real time," as the saying goes, have often succeeded in reducing the problem. But they have certainly not solved it.

Third, accuracy. Accurate information is often hard enough to obtain in peace. How much more so in war! The advent of modern precision-guided weapons, far from easing the problem, have made it much more difficult still. World War II bombers only needed to know where some German or Japanese city was located. Their twenty-first-century successors need to know the location of much smaller targets such as headquarters, bridges, power plants, and, in counterinsurgency, individual people.

Fourth, reliability. Some sources may be more reliable than others. However, there probably is not one that is completely so. Some sources may be reliable for certain purposes and at certain times, but not for and at others. The only way to deal with these problems is to use as many different sources as possible, understand the capabilities and limitations of each separately, examine the truthfulness of each piece of information they bring, and compare them. All in order to "make sense" of them and answer the questions we are interested in—assuming, that is, we know what those questions are! That process is commonly known as interpretation.

Early in the twenty-first century, the greatest obstacle facing interpretation may well be a vast surplus of data. In the Gulf conflict in 1990–1, so much information did U.S. satellites gather that the ground forces were overwhelmed by it. These problems explain the rise of data mining, i.e. the use of fast computers to go through vast amounts of information in the hope of discovering patterns in it. Yet such techniques can be circumvented or fooled. Especially when it comes to answering specific questions, most of the process can only be carried out by humans. That makes it both time-consuming and highly subjective. Personality traits, prejudices, biases, loves, and hates all enter into the equation and influence it. The person who can interpret information without anger and without favoritism, as the Roman historian Tacitus (AD c.56–after 117) famously put it, has yet to be born.[7] Sorting out these factors and assembling a sound assessment presents a formidable problem that has never been adequately solved. Nor, as long as war is waged by humans, is it likely to be solved in the future.

The four problems are interconnected. Incomplete intelligence is unreliable by definition. Obtaining reliable, accurate, and complete intelligence may take a long time. As with logistics, waiting until all the pieces of the puzzle are in place may mean waiting forever. Timeliness will depend partly on the technical means to hand, partly on other factors such as organization. The side able to obtain information, interpret it, and act on it faster than the other will acquire a highly important, quite possibly decisive advantage.

Moreover, intelligence is a two-sided business. Each side seeks to form a correct picture of the other while simultaneously denying him information and/or presenting him false information about himself. Thus the paramount tools of intelligence are secrecy and deception. Both will make each operation and each attempt at coordination more cumbersome than usual. As is said to have happened prior to the September 2001 attack on the Twin Towers, they may also create a situation where the right hand does not know what the left one is doing. Those are problems that must be dealt with and solved.

The outcome is a dynamic process. Thinking so and so, I must make him think that I think that he thinks that I think that he thinks...all this, while taking his intentions, capabilities, and prejudices into account. In theory, the outcome will be an endless series of mirror images. In practice, whereas strong chess players in the opening stages of their games can look ahead ten moves and more, on few if any documented occasions did commanders anticipate more than two. Often even preempting the opponent by a single move is quite an achievement. Too much playing with mirrors while trying to guess the opponent's intentions may erase the line between truth and falsehood not only in the enemy's mind but in one's own. The outcome is confusion and impotence. That is one reason why the equations of game-theorists rarely appeal to practically minded soldiers.

Secrecy, bluff, feints, and deception are used to obtain surprise. Of all the means that lead to victory, surprise is the most important. Especially at the lower levels where a single blow is most effective. In theory good intelligence, meaning such as follows the enemy's intentions and capabilities and penetrates his mind, should be able to prevent surprise. In practice, experience shows that doing so is extremely difficult, often impossible.

Fear of surprise often leads to a whole series of measures designed to survive it and reduce its impact. They start with dispersion, camouflage, hardening, and fortification and pass through the creation of redundancy and deployment in depth all the way to watchfulness and exercises aimed at training people to cope with the unexpected. Each of these measures is important and necessary. However, only a fool will believe they are fail-proof. Often the fog of war is only lifted when the enemy is already on us—or we are on him.

Finally, information on its own is like a bouquet of flowers. Nice to have, but useless. It is valuable only if and when acted upon. In other words, if it is integrated with operations. Those responsible for intelligence must have free access to commanders. Commanders must be able to look over the shoulders of those in charge of intelligence. As the case of drone-operators illustrates very well, below a certain level the two kinds of personnel must merge. The closer to the enemy we get, and the faster the reaction times required, the more important it is that the chain of command be organized accordingly.

VI

On Strategy

1. Strategy's Toolbox

Focusing on war proper, and ignoring all the other contexts in which the term is used, strategy can have two different meanings. The first is associated with Clausewitz. It refers to the major operations that bridge the gap between the nonviolent contest known as politics and the violent, but relatively small-scale, level known as tactics, where the actual fighting takes place. The other is associated with Sun Tzu (who, however, never uses the word). It is the art of conducting a contest between sentient opponents. Here we shall use the term in its second sense.

Before proceeding, it is necessary to draw attention to two elementary facts. First, strategy is measured by results. The most elegantly laid plans, the most powerful advances, the most beautiful maneuvers, and the most sophisticated stratagems are useless if they end in defeat. Second, war is a contest whose essence is the *interaction* between sentient opponents. This implies that a single powerful blow, delivered at the outset, which leaves the enemy incapable of resisting is not war; rather, what such a blow does is to render war unnecessary.

Depending on geography, economics, technology, culture, and similar factors, some people may habitually prefer some strategies to others. Tribal societies typically rely on the raid and the ambush. By some accounts, Westerners have always preferred concentration, frontal battle, shock attacks, and the breakthrough. Not so Easterners, especially those living on the steppe, who went for swarming, envelopment, and encirclement. Some societies relied on fortresses, others disdained them. The kind of weapons used, and the environment in which the struggle is fought (land, sea, air, outer space, or cyberspace) matter greatly in choosing the right strategy and implementing it. So do different kinds of terrain.

Nevertheless, at bottom the principles of strategy—pure strategy, as Clausewitz, at one point, calls it—are immutable. Starting with comparatively minor skirmishes, they apply to battles and even entire campaigns. Beyond a certain minimum, the size of the forces involved does not matter either. That is why, in this study, we shall refrain from drawing the usual distinction between strategy and tactics. Given that the former term seems to be taking over from the latter, doing so is in accordance with common usage.

Any strategic plan must start by allocating resources to targets and overseeing their coordination as they go into action. That, however, is just the beginning. If strategy involves anything, it is coping with the enemy's shifting intentions and moves. Mike Tyson, several times world heavyweight boxing champion during the 1980s and 1990s, is quoted as saying that everybody has a plan till they are punched in the mouth. That is why, as Moltke, Sr. wrote, no plan can survive the first clash with the enemy. Beyond that, strategy is simply "a system of expedients."[1] Move and countermove, move and countermove. Often he who can move faster than the enemy can react wins.

The principles of strategy are simple and few in number. What difficulties to the understanding it presents result from gaps in our intelligence; in other words, uncertainty as to what the future may bring and the opponent may do. An intelligent opponent is going to be unpredictable, an unintelligent one perhaps even more so. Other things being equal, the larger the war, the more numerous and complex the weapons, and the more complicated the environment in which it is fought, the greater the difficulties. To quote Napoleon, at the highest level strategy requires intellectual abilities not inferior to those of Newton.[2]

Forming the capstone of war, the purpose of strategy is to make the enemy comply with our will. As Sun Tzu, says, the highest triumph is won without fighting, purely by influencing the enemy's mind.[3] Ideally he will not understand he is lost until it is too late. Doing so requires almost superhuman insight and foresight. It also requires supreme flexibility and a well-nigh miraculous ability to overcome friction. So sublime is the ideal that it can rarely be attained. The target may be the mind; but normally it can be reached only by way of the body. Hence every plan must start by identifying an immediate *physical* objective. Such as inflicting death and destruction; or enveloping and encircling the opponent's forces; or occupying territory; or capturing key points; or paralyzing his command and control system; and the like.

The choice of objectives should be governed by two basic considerations. The first is the contribution they can make to victory. The second is feasibility. The latter is governed not merely by the capabilities of one side but also, and often mainly, by the other side's resistance. A modern tank can drive at 40 miles an hour and cover 300 miles before refueling. In practice, cases when an armored force covered over 30 miles a day for more than a few days on end are rare. In 2003 the Americans took three weeks to reach Baghdad. Thus their average rate of progress was 5 percent of the theoretical maximum! Most of the gap was due to Iraqi resistance, however light. The rest may be put down to the Americans' prudence and to friction.

At any one stage in the campaign, an ideal objective is so vital that attacking, destroying, or capturing it will lead to a general collapse. It should also be vulnerable and within our capabilities. It is against such an objective that what the Germans called the *Schwerpunkt*, the center of gravity, should be directed. A good example is the line between two formations, large or small. In October 1973 the Israel Defense Force's discovery of the seam between the Egyptian 2nd and 3rd Armies enabled it to cross the Suez Canal and turn the war around. But the opponent is not stupid. He will do what he can to avoid vulnerabilities and defend what he sees as the most important points. Trying to find them may turn out to be chasing a will-o'-the-wisp.

The objective having been chosen, one must examine whether, given the available means and the opponent's anticipated resistance, it is within reach. Next, one must retrace one's steps and determine whether it really contributes to victory. Ideally the desirable and the possible should coincide. In practice, owing to the limitations of intelligence and our inability to foresee the future, such coincidence is rare. Of all the errors strategists commit this one is probably the most common. For example, when the Germans invaded the Soviet Union in 1941 they believed the Red Army had 200 divisions, whereas in fact there were 360. Consequently the Wehrmacht did not succeed in crushing its opponent. It did not capture either Moscow or Leningrad and failed to break its opponent's will.

The opposite error is to choose the objective on the basis of our ability to achieve it regardless of the contribution it can make towards victory. An excellent example is the Japanese attack on Pearl Harbor. Both planning and execution were near perfect (though an intelligence failure meant that the American carriers, representing the main force of the Pacific Fleet, were not in port at the time it was attacked). However, the policymakers in Tokyo assumed that their enemies were decadent, reluctant to mobilize, fight, and sacrifice. They hardly asked themselves how the blow would affect America's will. Had they done so, perhaps they would never have dared seize the tiger by the tail.

The objective having been selected, detailed planning can start. It includes allocating forces and resources, dividing the mission between them, building a system of command and control, selecting the routes to be taken, and briefing. All these are essential for success. Often they are extremely complicated to perform. To adduce another example, how does one deploy half a million troops with all their equipment in the desert along the Saudi–Iraqi border far west of Kuwait, keep them supplied, and ensure that each commander and each unit know their missions; all without alerting the opponent?

The best strategy is always to be very strong, both quantitatively and qualitatively. The two things are inversely related. Normally increasing quantity will cause a decline in quality, and vice versa. The World War I British mathematician Frederick Lanchester argued that, to balance a quantitative advantage of 2:1, a qualitative one of

4:1 is needed.[4] If the quantitative advantage is 3:1 the qualitative gap should be 9:1, and so on. Reality is more complicated. So much so that, some very limited cases apart, all attempts to reduce it to mathematical formulae have failed. Finding the correct balance between the two things, both in general and when facing a specific opponent under specific circumstances, is extraordinarily hard.

Assuming threats and a show of force are not enough, several methods for breaking the enemy's will present themselves. A short war and a long war; offense and defense; annihilation and attrition. All can be combined in different ways, at the different levels on which war is fought, simultaneously or in succession. As a rule, the fastest way to victory is to combine a short war with an annihilating offense. Normally that is the choice of the strong. Napoleon, who liked to lead the larger battalions, used to say that, unlike other generals, he always focused all his efforts on the opponent's jugular.[5] By forcing him to defend it, he created an opportunity to attack him and crush him.

Psychologically speaking, going on the attack is often easier than having to wait for it to come. Another advantage the attacker enjoys is the initiative. It permits him to dictate the time and place of the campaign, determine the amount of resources he wants to commit, get in the first blow, shape the campaign (or, at a lower level, the battle), and, in general, call the tune. Unable to do the same, the defender must be content with reacting. For example, originally fighter pilots used to identify one another by sight alone. This enabled the attacker to use sun and clouds in his favor, whereas the defender at first found himself in an inferior position. As a result, countless pilots were shot down before they ever saw the enemy.

Cases when the attacker went for a long war and attrition are fairly rare. One prominent example was the Battle of Verdun in 1916. On that occasion the German commander, General Erich von Falkenhayn, hoped to use his artillery to bleed the French Army until it literally had no troops left. Another example is the so-called "War of Attrition" Egypt waged against Israel in 1969–70. In both cases, incidentally, the outcome was failure. The defenders' will remained unbroken. They did not retreat, they did not run, and they did not surrender. Having suffered heavy casualties, the attackers ended by suspending the offensive without achieving their objective.

Since the defender will find it easier to find cover in the terrain or to use fortification, normally he will need fewer troops than the attacker does. Hence the first choice of the weaker party may be to adopt a defensive posture. Not only can he make better use of cover, but his lines of communication remain steady. He does not have to detach troops to occupy and garrison conquered territory either. As long as he does not lose, he wins. That is why Clausewitz says the defense is the stronger form of war.

A defender who tries to be strong everywhere will end up weak everywhere. Should he stay inactive his forces will be demoralized, as happened to the French soldiers manning the Maginot Line in 1940. Should he try to resist, he will be gradually worn down. These considerations may compel the smaller party to gird its loins, take the offensive, and try to crush the opponent before the latter can bring his numerical superiority to bear. Frederick II in his instructions to his generals strongly advocated such a course. The famous Schlieffen Plan around the turn of the twentieth century mandated it. Another excellent example is the 1967 Israeli offensive against Egypt, Jordan, and Syria. However, this method is risky. Of the above, only the last succeeded. The others led to long struggles of attrition, precisely what their authors had hoped to avoid. Neither the North Korean attempt to defeat the South in 1950, nor the Pakistani one to defeat India in 1965, nor the Iraqi one to defeat Iran in 1981 succeeded.

At times one may take advantage of empty spaces in order to launch a large-scale offensive while still enabling the troops to fight defensively if and when they run into the enemy. This was the course Moltke, Sr. in particular recommended.[6] The number of possible combinations is practically endless; normally the larger the forces, and the greater the scale on which operations are conducted, the more true this is. Furthermore, most armed forces, and modern ones more than most, are far from homogeneous. They are large organizations made up of many different units, many of which are organized, equipped, and trained in different ways. Some rely primarily on firepower, others on mobility. Some are meant for offense, others for defense. Some are suitable for fighting in open terrain. Others specialize in mountain warfare, or air warfare, or sea warfare, or whatever.

Each unit can fight on its own. The trick is to combine them all. Combined arms means integrating various units and weapons in such a way that each will bring its own strengths to bear while at the same time having its weaknesses covered by the rest. By doing so, a whole can be created that is much greater than the sum of its parts. Two and two does not equal four but five, or six, or seven. Such a force should be capable of coping with various threats as required.

One way to use combined arms is to put the opponent on the horns of a dilemma to which there is no solution. For centuries on end, European commanders did so by using cavalry to force the opposing infantry to form dense formations known as squares; the reason being that only infantrymen so arrayed can face the faster, more powerful horsemen. Once the squares had been formed they could be targeted by artillery and forced to disperse. Thus the opponent was damned both if he did and if he did not. Similarly, in World War I the gunners on both sides often fired a mixture of high explosive and gas. The former would force the opponent to take shelter. The latter, being heavier than air, compelled him to abandon it.

Closely related to, but nevertheless different from, combined arms is orchestration. The simplest way to explain orchestration is to use an analogy with chess. Even an average chess player understands what is meant by this: namely, using the various pieces in such a way that each will serve not a single purpose but several at once. The white pawn on line 6 or 7 covers its king but also threatens to be queened (i.e., being turned into a queen on reaching the opponent's base line). A bishop, correctly positioned, can dominate certain squares, threaten both the opponent's knight and his rook, and block a possible attack on its king. The larger the number of purposes served by each piece separately and by all combined, the better the play. As the game develops the purpose of each piece keeps changing, but the principle remains.

Imagine an army making its way towards some important objective. Unit A forms a reserve, but is also ready to defend against an opponent coming from another direction. Unit B carries out a diversionary attack but is also in charge of protecting the army's communications. Units C and D provide mutual cover, while also threatening the opponent. The number of possibilities is infinite. This is the field in which Napoleon was the unsurpassed master. He deployed his corps, normally eight in number, like the waving arms of an octopus, engaged the enemy, forced him to stay still, and closed upon him.

2. Strategy in Action

Let us return to strategy as a two-sided struggle where the moves of each side reflect those of the other. Such being the case, the way it works, and should work, is best described by means of pairs of opposites. The number of pairs is almost infinite. Many are linked and overlapping; trying to discuss them all will lead to endless repetition. Those listed below are among the most important. Yet they are meant to illustrate the subject, not to exhaust it.

A. *Maintenance of Aim versus Flexibility*

"Maintenance of aim" is one of the cardinal principles of strategy. The determination to overcome obstacles, take losses if necessary, and proceed towards one's goal is essential for success. However, one may go too far in this respect. There do exist situations and moments when a new course must be set.

Since it is a question of looking into the future, and since hope dies last, identifying those situations and moments is very hard. Yet doing so is essential, or else maintenance of aim may turn into hitting one's head against a brick wall. The Italians on the river Isonzo in 1915–17 attacked the Austrians no fewer than eleven times. They lost every battle and took hundreds of thousands of casualties. Worse, the attacks exhausted the army. They opened the way to the October 1917 Austrian counteroffensive at Caporetto which almost knocked Italy out of the war.

Shifting now one way, now another, war gives the advantage to commanders who can seize any opportunity that comes their way. Their plans must have not one branch but several. Reserves, consisting of material resources, or unengaged forces, or both, are vital. They allow the attacker to change the objective of the attack, its place, its methods, and its force. Conversely, they enable the defender to reinforce weak spots or to counterattack. In both cases, timing is everything. The attacker must throw in his reserve when the enemy starts to weaken; the defender, when the attack reaches its culminating point.

Normally reserves should amount to one tenth to one third of the force. Less is dangerous, more wasteful. Lacking reserves, a commander will find it much harder to influence the campaign. He may be able to do so, if at all, only by taking a risk and moving forces from one sector or theater to another. When Winston Churchill, in the midst of the 1940 German invasion of France, asked the French High Command where its reserves were and was told that there were none, he immediately knew everything was lost.[7]

The supreme form of flexibility is to fool the opponent by voluntarily doing what he wants before he forces one to. The Mongol false retreat, used to lure the opponent forward, surround him, and annihilate him, was famous. Starting with the Roman general Crassus in 53 BC, countless commanders fell victim to it, as King Harold of England probably did at the Battle of Hastings in AD 1066. Another example is the 1917 German retreat to the Hindenburg Line. It forced the Allies to invest time and resources in pushing their forces forward and building a new infrastructure on the Western Front, while saving the Germans seventeen divisions to do with as they pleased.

A voluntary retreat will preserve morale and prestige in a way that a forced one will not. Yet flexibility, like its opposite maintenance of aim, must have its limits. Too much flexibility may mean giving up the initiative. Contradictory orders will cause confusion and demoralization. Either way, the outcome can be defeat.

B. *Husbanding Force versus Sacrificing It*

Supposing a single blow does not end the war, one of the strategist's main tasks is to husband his forces. Doing so is especially important in a war of attrition where the last available division, kept in reserve, may well decide the issue. Another method is to stay on the defense, which normally requires fewer forces than the offense does. That is what the Germans on the Western Front did from 1916 to the spring of 1918. As a result, they invested considerably less in killing each enemy soldier than the Allies did.

As so often, the principle is easier to explain than to implement. A force that is being husbanded can only be used within strict limits. This may reach the point of pure passivity. A perfect example is the French Air Force in 1940. Kept in reserve—the French High Command believed the war would be a long one and that the last man and the last aircraft would decide the issue—it hardly tried to interfere with the German steamroller. As a result, when the campaign ended the number of combat-ready French aircraft was actually higher than it had been at the beginning!

Rarely, if ever, have wars been won by pure defense. Even the famous Roman commander Fabius Cunctator ("the delayer") ended by losing his job to Scipio Africanus. The latter, a younger and more active man, took the offensive against Hannibal and defeated him. Even the Red Army, which, attempting to husband its forces in 1942, gave up tens of thousands of square miles of territory, ended by halting at Stalingrad. There it stood, fought, and took horrendous losses before counterattacking.

Moreover, sacrificing forces may be absolutely necessary. A commander on the defensive may sacrifice some of his troops so as to gain time and save the rest. As in ordering a surrounded fortress to hold out at any cost; or having a bridge blown up in spite of the fact that some of his troops are still on the other side. To save his ship a captain, knowing full well that part of the crew will be lost, may order the watertight doors closed.

Sacrificing force may also be part of an offensive, when it serves to mislead the opponent, tempt him into making the wrong moves, and lead him into a trap. What all these cases have in common is that human lives are deliberately sacrificed. A commander who cannot make himself sacrifice some of his forces when necessary, or who is prohibited from doing so by his superiors and the society he is fighting for, will not be able to fight and win a war.

C. Concentration versus Dispersion

To quote Clausewitz, the first rule of strategy is to be as strong as possible, first in general and then at the decisive point.[8] The second, which follows from the first, is to concentrate one's forces against the right enemy, at the right place, at the right time, with the right objective in mind. A force that does not participate in accomplishing the principal task of breaking the enemy's will is a force wasted. If it has to be kept supplied and secured it may actually do harm.

But there are also problems. First, a concentrated force will necessarily leave other fronts and/or the rest of the front exposed. Doing so may be risky indeed; even the Israelis, when attacking the Egyptian Air Force in June 1967, only used about 95 percent of their available combat aircraft. Second, concentrating a force may mean having to do without a reserve—with all the attendant risks.

Finally, a concentrated force is easier to discover than a dispersed one. Its intentions are also easier to gauge. As a result, it is relatively easy to foil. That can be done by opposing it with an equal or superior force; or by a geographic movement that will turn its move into a blow in the air; or by threatening it from another direction. All may compel a concentrated force to disperse. Having dispersed, however, it may be unable to accomplish its objective and risks being picked off in piecemeal fashion, bit by bit.

D. Battle versus Maneuver

Tribal warriors—meaning, all warriors until the onset of the agricultural revolution sometime around 10,000 BC—relied almost exclusively on the raid, the ambush, and the skirmish. More advanced societies with larger armed forces at their disposal were able to fight regular battles, *battailles rangées*, as the French phrase goes. Such battles pitted the main forces at the disposal of each side against each other in a single "field." They did so, however, not by stealth and surprise but in the open, often after prolonged standoffs that might last hours, days, or weeks.

Machiavelli wondered whether it is better to fight like a lion or like a fox.[9] He was by no means the first or the last to do so. Battles were always risky affairs. A really successful battle may end a war almost at once. As, for example, when the Romans beat the Macedonians at Pydna in 168 BC and also when the Prussians defeated the Austrians at Koeniggraetz in 1866.

A battle lost could lead to the same result, only in reverse. For this reason, but also in order to exploit the advantages of coming at the enemy from an unexpected direction, leveraging him, unbalancing him, and chopping off his limbs before focusing on his trunk, many commanders have always preferred maneuver. The Marshal de Saxe (1696–1750), one of the more successful commanders of Louis XV and a fine military writer, claimed a good general might wage war all his life without having to do battle.[10] Yet he himself fought at least three, including the one at Fontenoy (1744) which enabled France to end the War of the Austrian Succession as one of the victors.

Battle and maneuvers are opposites. But they are also complementary. A good commander should use them both as circumstances may dictate, making each reinforce the other. That said, it is important to add that, since 1945 or so, the role of battle in war, as traditionally understood, has been declining. The reason for this is the vast firepower of modern weapons. It compels armies to disperse and no longer allows them to concentrate their main forces, let alone all their forces, at a single spot. That, however, does not apply to smaller formations operating at lower levels. With them, the choice between battle, or perhaps one should say fighting, on the one hand and maneuver on the other remains as relevant as ever.

E. Direct versus Indirect Approach

The shortest way from A to B is always a straight line. However, normally taking the direct approach will cause the opponent to block it. The outcome, a frontal clash, is likely to be bloody and indecisive. Accordingly one should avoid the direct approach, make indirect and unexpected moves, and start by attacking the places where the enemy is weak in order to cut off his arms and legs, so to speak. Thus the shortest line becomes the longest, and the other way around; the indirect approach becomes direct, and vice versa.

Coming out of World War I, during which both sides repeatedly paid a very high price for their attempts at direct attack, the British military historian and commentator Basil Liddell Hart developed this idea into a doctrine and a theory.[11] At its heart was the claim that the indirect approach was the most important tool of strategy and that it was (almost) the only way to win victory in war. A good example is Napoleon's crossing of the Alps in the winter of 1798–9, which led him straight into the Austrian rear and from there to the Battle of Marengo. Other examples are not lacking.

F. Breakthrough versus Envelopment

As a rule, the fastest road to victory is to break through the opponent's center, causing his force to disintegrate into fragments that can be overcome separately, if indeed they have to be overcome at all. While many victories have been won in this way, this method suffers from two disadvantages. First, assuming the opponent is on guard, achieving a breakthrough is necessarily hard and is likely to meet with strong resistance. Second, advancing into the opponent's territory inevitably means exposing oneself and putting one's neck into a noose. Israel's crossing of the Suez Canal back in October 1973, during which its troops were almost cut off by the Egyptians, provides an excellent example of such a situation.

The other possibility is envelopment that may lead to complete encirclement. Both envelopment (outflanking) and encirclement create a situation whereby one side attacks the other from an unexpected direction and compels him to redeploy. Both the opponent's line of communications and his line of retreat may be cut, forcing him either to fight under unfavorable circumstances or to surrender. Furthermore, to quote Ariel Sharon again, nothing so terrifies soldiers as do enemy troops suddenly appearing in their rear.

As always, there are also disadvantages. First, whoever outflanks is automatically outflanked. Second, encirclement may well force one to divide one's forces. That in turn demands numerical superiority, or else the ring will be too thin and easily broken. Apparently, fear lest this might happen caused the Allies to refrain from drawing the noose around the Germans at Falaise in 1944. Whether, in any specific case, the advantages are greater than the disadvantages, or vice versa, depends on the specific circumstances of the situation.

G. Advance versus Retreat

With few exceptions—by far the most important of which is the surrender of Japan in 1945—victory, partial or complete, is achieved by physically advancing into the enemy's territory. Next, with or without a battle having been fought, key objectives must be captured, such as military bases, geographical and topographical points, communications nodes, natural resources, industrial plant, cities, and various symbolic objectives. Occupation will augment the resources and raise the morale of one side and do the opposite to the other. It will also help bring about the moment when the opponent's will is broken.

But advancing into hostile territory does not come without a price. The invader's communications are extended, compelling him to detach forces so as to garrison his conquests. By the time Napoleon, advancing on Moscow, reached the city he had lost or left behind two thirds of his troops, causing his initial numerical superiority to vanish. Another problem is the occupied population. The more time passes, the more likely it is to turn against the invader. An advance is like water leaving an overturned bucket. At first the flow is strong and fast; but the further it goes, the weaker it becomes until it stops altogether.

The situation of the retreating side is just the opposite. A retreat, unless it is voluntary and unless its advantages can be clearly explained, will damage morale and can occasion a panic. The closer the pursuit, the more likely this is to happen. Withdrawal causes resources to be lost. On the other hand, the more one retreats, the shorter one's communications become and the closer one gets to the centers of one's power. One's very weakness may paradoxically increase one's strength.

In the end there are two possibilities. A resource too precious to be surrendered, a frontier, or a natural obstacle may force the retreating side to stop his withdrawal and give battle. Or else he may simply break down and give up. Unless either of these things takes place, the outcome will be a culminating point in which the two sides' roles are reversed. That is what happened to the Germans at the Marne in 1914; and again at the gates of Moscow twenty-seven years later.

The decisive factor is time. The attacker only has limited time to achieve his objectives. Failing to do so, he will be tied down. Debilitated, he may end by being defeated. For the defender, the situation is just the opposite. As long as he is not defeated, he wins.

H. Strength versus Weakness

To speak with Sun Tzu, a good strategist uses rocks to break eggs, and eggs—to camouflage or neutralize rocks. He concentrates strength against weakness and uses weakness at one spot to concentrate strength at another. He puts being against non-being; fullness against emptiness, and emptiness against fullness; the unexpected against the expected, the expected against the unexpected. Like fireflies, they should be constantly switching places. All this, without the strategist losing sight of his objective or allowing things to escape his control.

When it comes to the conduct of strategy, the weaknesses on each side, meaning the things each side cannot or does not want to do, are at least as important as its strengths. One side must set a trap and the other—march into it. An excellent example is the Battle of Cannae in 216 BC. On one hand were the Romans, accustomed to advance straight forward so as to bring their superior discipline and fighting power to bear. On the other was the Carthaginian commander Hannibal. He used this Roman propensity to advance straight forward to lead them by the nose, inducing them to advance into his retreating array, enveloping them, and killing 70,000 of them in a single day. It was the worst defeat Rome suffered in an entire millennium. What made it possible was the fact that the strengths of one side fitted exactly, almost perversely, into the weaknesses of the other. And the other way around.

Such a fit is extremely rare. Accident apart, such a situation can usually be brought about only by deception. Deception in turn must be based on a thorough understanding of the opponent. In a sense, indeed, no attempt at deception can succeed unless the other party is willing to deceive himself. To quote the German chief of staff, Field Marshal Alfred von Schlieffen (served 1893–1905), a really great victory requires that the opposing parties cooperate, each in his own way.[12]

3. The Ordinary and the Extraordinary

Maintenance of aim and flexibility; husbanding force and sacrificing it; concentration and dispersion; battle and maneuver; direct and indirect approach; advance and retreat; breakthrough and envelopment; strength versus weakness; advance versus retreat; these are just a few of the many pairs of opposites that characterize strategy. Inside each pair, each method relates to the other as light does to darkness, the ordinary to the extraordinary. They complement each other, but can only exist *at the expense* of the other. Other things being equal, the forces sent to create a diversion will be unable to participate in the main push. Those used to envelop the enemy will not be able to assault his front or form part of the reserve. In other words, to any given course of action there is always a cost. As a struggle unfolds, its character will be governed by the way both sides navigate between these opposites.

Aristotle believed that people and groups going through life should find the golden mean between such opposites. Neither too much nor too little was his motto. His approach makes a lot of sense, all the more so because we are unable to forecast the future. As a result, it is best to take a balanced approach while preparing for every contingency.

Applied to military strategy, the advice is problematic. To repeat, a good strategist uses rocks to break eggs, and eggs—to camouflage or neutralize rocks. Following the golden mean implies going for a frontal clash, face to face, without any attempt to deceive the opponent, or circumvent him, or leverage him, or maneuver him into an untenable position. Applying this approach against a mediocre opponent, the result is apt to be equally mediocre. Applying it against one who is more than mediocre, the outcome is likely to be defeat.

Thus the correct conduct of strategy requires not just the ability to foresee, and cope with, all possible threats. It also requires the willingness to create *asymmetric* situations. Now the situation calls for one method, now for its opposite. They should take each other's place according to a well-defined plan, yet without any obvious order that the enemy can discern and exploit. Following such a course of action can be risky. A commander who, having taken such a risk, fails in his endeavor is almost certain to be deposed. Under some regimes he may be executed. However, unless the odds are greatly stacked in his favor to begin with, risk-taking alone can lead to good results.

Again, this is easier said than done. Any successful strategy must start by studying the opponent, getting to know him, and adapting one's methods to fit his *modus operandi*, even his nature. Since the effort is mutual, the more time passes the more both sides will change, adapt themselves to fighting the specific opponent at hand, and become alike.

Therefore, given enough time, an asymmetric contest will become symmetric. As the process unfolds, the weak may become strong and the strong, weak. A football team that is only allowed to play a much weaker opponent will lose its edge. The same is true of a military.

A look at the American campaign in Afghanistan, which was easy at first but later deteriorated into bloody stalemate, will confirm the need to avoid such a situation at almost any cost. Other examples are not lacking.

Since the essence of strategy is deception, and supposing the opponent is no fool, repeating the same move, even the most successful one, is the highway to disaster. Notwithstanding the fact that, on occasion, such repetition itself may take the opponent by surprise. Accordingly, the first prerequisite for the successful conduct of strategy consists of a thorough understanding of the opponent including, above all, his expectations of us.

The above rules, to the extent that they deserve the name, can be mastered by both sides. That is why acting within the existing, well-known parameters is likely to be a recipe not for success but for failure. What is really needed is *breaking* the rules and putting others in their place. Light must be turned into darkness, or the other way around. The impossible must become the possible. An excellent case in point is the Syrian and Egyptian decision to start the 1973 October War in spite of their opponents' command of the air. That was something neither the Israel Defense Forces nor many other armed forces around the world had ever considered possible. As General Douglas MacArthur once said, it is for breaking the ordinary rules of strategy that great commanders are remembered.[13] In the end, only one word can express the necessary quality: genius.

All this means that plans must be laid down, intelligence gathered, preparations made, and the war conducted according to the rules of the art—and then some. Yet this is not to say strategy is omnipotent. Partly because of the role played by moral factors, partly because it is based on expectations which are always subjective, except at the lowest level any attempt to reduce it to algorithms is foredoomed. However good our calculations, they can never yield more than probabilities. It is a question involving not just the intellect but also, perhaps mainly, intuition and imagination.

Finally, even the best-planned, best-informed, and best-prepared strategy, executed by a genius, may be disrupted by chance events. Conversely, chance may crown the most incompetent strategy with success. In the long run, if a long run there is, Dame Fortune tends to favor the competent. However, as Machiavelli wrote, she is fickle and her actions can never be foreseen.[14]

VII

War at Sea

1. Maritime Facts of Life

Surprisingly, neither Sun Tzu nor Clausewitz in their respective works has a word to say about maritime war. Jomini does devote a chapter to the subject; but he only treats it as an aid to land warfare. He has a long list of "descents," made by this commander or that, with or without success. Jomini was a native of Switzerland, a landlocked country. Though he served in several other armies, he never renounced his citizenship. Is that why he shows no understanding whatsoever of the sea as a different environment where different laws hold? At any rate maritime war, let alone command of the sea, its importance and the ways to attain, retain, and disrupt it, are barely mentioned.

The naval theorists Alfred Mahan and Julian Corbett both tried to place war at sea within war as a whole. Mahan took Jomini as his model. Corbett preferred Clausewitz, emphasizing his distinction between "absolute" and "limited" war and the way that naval power could be used in waging the latter. However, given their subject, both put much more emphasis on the tail, i.e. maritime strategy, than on the dog, war, of which it is a part and by which it is wagged.

Whatever the reasons, a first-class, up-to-date study, incorporating both aspects of war and pointing out the similarities and the differences between war on land and war at sea and the ways in which they interact, remains to be written. And yet, as thirteenth-century BC reliefs showing the Philistines trying to invade Egypt by sea prove, the two have always been linked. Had the Achaeans not had a thousand ships, not even the face of the most beautiful woman on earth could have lured them to Troy. Nor would the Athenians have been able to turn back the Persian invasion at the Battle of Salamis in 480 BC.

Land and sea present very different environments. These differences go far to govern the kinds of war that can and cannot, have and have not been, waged in them. Land war has always had to consider the surrounding population as well as the available resources and products. To exploit the human and material resources of a district or country it was necessary to conquer it first. Often that had to be done almost inch by inch—and dead, injured, or captured enemy person by dead, injured, or captured enemy person. Only when the process was more-or-less complete could any benefits accrue to the conqueror.

The situation at sea was and is entirely different. There are no people and few resources to exploit. During most of history the latter consisted almost exclusively of fish. Fish was sometimes a favorite dish. But rarely was it so important to a country or people that they could be brought to their knees by being deprived of it. Recent technological developments, by extending man's reach from the surface of the sea to its bottom, are changing this situation. Offshore resources such as oil and gas are critical for the economies of many countries. They may yet be joined by various minerals, huge quantities of which are found on or under the bottom of the sea. Even so, since consumers don't live in the sea, those resources can only be exploited if they are first brought to the land where they are distributed and consumed. In other words, communications are everything.

Maritime communications are often known as sea lanes. To avoid confusion, here we shall use the latter term throughout. Both on land and at sea, communications/lanes link bases with destinations. Through them flows a constant stream of reinforcements and supplies; should access to base or port be cut, then normally it is only a question of time before the forces they serve will be forced to retreat or surrender. Sea lanes resemble lines of communication in that there are various choke points such as capes, straits, and canals. These points often enable their owners to dictate the enemy's strategic moves. The British Empire in its heyday controlled the Straits of Dover, the Straits between Scotland and Norway (World War I), and those between Norway and Iceland (World War II). It also held those of Gibraltar, Sicily, Aden, Hormuz, Magellan and Malacca, as well as the Cape of Good Hope and the Suez Canal. But for them, it would scarcely have survived.

Some friends, some enemies, and some locations can be reached only by sea. Had Britain and Japan not been islands their history, military history included, would have been utterly different. The same applies to the United States, a "global" island protected by the Atlantic on one side and the Pacific on the other. Speaking to the Reichstag in April 1939, Hitler said that the idea of invading the U.S. could only originate in a disturbed military imagination.[1] Back in 1985, incidentally, that did not prevent one American company from making a movie about just such an invasion—by Cuba!

What is more, transportation by water has always been easier and cheaper than by land. Not by accident did all the great ancient civilizations, Far Eastern, Middle Eastern, and Mediterranean, rise either along rivers or near the coast. Landlocked peoples lagged far behind. Never was the advantage greater than during the so-called "Columbian" period from 1500 to 1850. On land the most advanced forms of transport were horse-drawn carts and carriages. Before the industrial revolution and the invention of steam their capabilities only increased very modestly. Even at the peak of their development the loads they could carry were limited to a few tons and their sustained speed to rather less than ten miles per hour. Ships, propelled by oars or sails, were no faster. Their dependence on the wind often made the duration of their voyages impossible to predict. However, their carrying capacity was much greater.

The introduction of railways around 1850 and of automobiles half a century later did something to close the gap. Nevertheless, ships remained the only means by which islands could be reached. Aircraft, which also date to the years around 1900, can overfly land and sea alike. However, their load-carrying capacity is limited. That explains why, though almost all the troops the U.S. and its allies sent to the Persian Gulf in 1990–1 and 2003 arrived there by air, 90 percent of the supplies did so by sea. Air transport plays a greater role in the war in landlocked Afghanistan, but only at an exorbitant cost. For all the technical development of automobiles, railways, and aircraft, had it not been for maritime transport, intercontinental trade would be reduced to a fraction of what it is.

The sea has neither terrain features nor surface vegetation. Their absence has always prevented ships from finding the kind of shelter so important in land warfare. Yet this very absence, as well as the fact that there are no inhabitants, also gave ships on the high seas greater freedom of movement. Often they were harder to track and nail down than armies. Nelson, vainly chasing the French fleet in the Mediterranean in 1798, provides an excellent example of this. As late as 1942 the Japanese at Midway had no idea where Admiral Fletcher's Carrier Task Force 17 was or that it was waiting for them. During the 1960s satellites started changing the situation, making ships and fleets all but impossible to hide. Today the only vessels that can still escape their attention to any considerable degree are submarines, a stealth weapon par excellence.

Finally, at sea as on land, climate and weather play an important role. Wind, currents and tides have often interfered with naval operations. The replacement of oars and sails by mechanical energy, as well as the construction of much larger vessels, somewhat reduced ships' dependence on them. But that does not mean they can be ignored. In particular, submariners and their opponents must keep in mind factors such as water pressure, temperature, and salinity.

Today as ever, meteorological and physical factors often make it advisable to follow certain routes and frequent certain regions at certain times while avoiding others. In 1274 and 1291 it was the weather which broke up the Mongols' landing in Japan. In 1588 it was the weather that first prevented the Armada from linking up with the Spanish Army in what is now Belgium and then completed its destruction. In 1944, had not the weather cleared up at the last moment, the

Normandy landings would have had to be postponed. These factors have always helped determined the principles of maritime strategy. They do so still.

2. Some Principles of Maritime Strategy

The land is the habitat on which all humans live. By contrast, the sea forms a strange, almost unnatural, environment for which evolution has not prepared us. Countless people of all ages never set their eyes on it. Some hardly even knew it existed. If humans are to use it for civilian or military purposes they must first learn how to master it and the vessels that navigate it. Doing so takes time and effort. That is why navalists have always put a heavy emphasis on professionalism, not seldom at the expense of coordination and cooperation with the other services ("jointness").

Given its absolute dependence on ships, war at sea has always been more capital-intensive than its land-bound equivalent. By and large, that applied even when ships were powered by oars. On the whole, the challenges war at sea presents to those who practice it—the responsibility, the uncertainty, the friction, the danger, etc.—are broadly similar to those prevailing on land. So are the methods of dealing with them, including training, education, organization, discipline, and leadership. The nature of the environment in which war is waged may affect its form. However, the qualities the warrior needs remain basically unchanged.

On land, technology (even if it consists only of sticks and stones) is absolutely essential for the conduct of operations. At sea, it is essential both for operations and for survival. On the high seas ships, later joined by submarines, have long been all-powerful. They go where they like and sink or capture whom they will. As the battles of Cadiz (1596), Aboukir (1798), Copenhagen (1807), Taranto and Dakar (both 1940) all showed, in port, unless it is a *very* strongly defended one, they are vulnerable. The same applies when they pass through, or try to operate in, narrows and straits. The more so because, starting around 1900, narrows and straits have often been mined.

Armies, as long as they are not involved in heavy fighting, can normally survive for at least some time without receiving fresh supplies by living off the country. The Austrian troops in Friuli, in northeastern Italy, in 1918 did so for a full year, albeit at a terrible cost both to the local population and to their own health. By contrast, fleets can hardly obtain any supplies from the sea. Until recently that included water. Resupply at sea from other ships is possible, but requires a constant flow of various kinds of auxiliary vessels that are both expensive and vulnerable. Maintenance and repair are only possible within strict limits. Thus practically everything has to be gathered together in port, loaded on board, and carried along. All this limits the time that various types of vessels can spend at sea as well as the distance they can travel before returning to the same port or looking for another. Nor is every port necessarily suited for every ship.

Many of the basics of strategy are equally valid in naval war. That includes the need to combine different kinds of ships and orchestrate them. Making due allowance for geography and the freedom of movement the high seas provide, it also includes most of the pairs of opposites that strategy uses and of which it consists. Since the most important use to which the sea is put is transport, from the time of ancient Athens on the first objective of navies has always been to keep their own sea lanes open. Assuming, as has been the case during most of Western naval history, that warships and merchantmen are not identical, basically there are but two ways of doing this. One is to use the former to escort the latter. As on land, this means giving up the initiative. As on land, too, trying to defend everything risks being left weak everywhere.

The alternative is to build a strong battle fleet based on the most powerful available vessels. Next, this fleet must either seek out the enemy or so position itself as to force him to fight. It must engage him and destroy him. Finally it must hunt down any survivors until they have been sunk or bottled up in port. Much the most important element of naval strategy is "command of the sea," *thalassokratia*, as the ancient Greeks called it. He who exercises it can do whatever he likes, wherever he likes, at any time he likes, on the high seas. More so than on land, in fact, for no army of occupation is needed. Made famous by Mahan, this approach remains the official doctrine of the U.S. Navy.

In practice, rarely does one side obtain complete command of the sea or hold it for very long. Much of the time it is disputed, as it was, for example, in both World Wars. When it exists it may be, and often is, local and temporary. Navies too weak to wrest command of the sea in battle can try to prevent the enemy from achieving it. One way of doing so is to mine the sea lines the enemy must use, especially near ports or in narrows. A second is to maintain a fleet in being that will represent a constant threat, thus limiting the enemy's freedom of action; a third, to engage on commerce-raiding. The third course implies reliance on dispersion and stealth. Speed, needed in order to overtake the enemy or get away from him in case he proves too strong, is also essential.

Historically *guerre de course* (or "war of the chase"), as it was sometimes known, has tended to focus on ports of departure and destination as well as choke points though which merchantmen must pass. But there were also cases when it extended into the high seas; at such times, the remedy was to group ships into convoys and provide them with escorts, close or covering. In 1911, just as Corbett was busily explaining why commerce raiding would probably be less effective in the future, submarines were introduced. Submarines could not achieve command of the sea. But they did prove a formidable instrument in denying it to others. In both World Wars German submarines came close to bringing Britain to its knees. The role of submarines in the Allied victory over Japan was equally great.

At one time in the bases of the Israeli Navy, one could read the following slogan: "never defeated was a people/that rules the sea" (in Hebrew, "people" rhymes with "sea"). The debate as to which of the two, land power or sea power, is more important and how they relate to each other goes back at least as far as Themistocles in the fifth century BC. It was he who convinced his fellow Athenians that they should build a fleet rather than try to resist the invading Persians on land. Both kinds of power have advantages and disadvantages. In essence, neither the advantages nor the disadvantages have changed from the earliest known times to the present. A fact, incidentally, that poses an interesting question as to the extent to which technological development can ever change the fundamentals of war.

To quote the English jurist and essayist Francis Bacon (1561–1626), in a conflict between two powers he who commands the sea has the immense advantage of being able to wage as much, or as little, war as he wants.[2] He can move troops and resources from one point to another. Or impose a blockade, close or distant, on the enemy so as to prevent him from moving *his* resources and *his* troops, either choking him to death or forcing him to come out and do battle; or "project power," as the phrase goes, to the places, and at the time, it is needed. Last but not least, a relatively small expeditionary force that holds the initiative can compel the defender to deploy far larger forces in an attempt to deter it or, should deterrence fail, throw it back into the sea.

As early as the Peloponnesian War (431–404 BC) the Athenians used their navy to hold their empire together, import the grain they needed, and raid the Peloponnese. The American North in 1861–5 blockaded the South, making a considerable contribution to the outcome. In 1944 the threat of an Allied invasion forced Hitler to maintain a million troops in the West—troops which, had they been available for the Eastern Front, might have turned the tide. For centuries on end, so systematically did the Admiralty in London apply these and similar methods that, collectively, they became known as "The British Way in Warfare."

Blockades can sometimes push even powerful continental countries to the brink of starvation, as happened to Germany during World War I. But they also involve tremendous wear and tear and are very expensive in terms of men and equipment. Nor are they always effective. A large country rich in agricultural products, raw materials, and industry should enjoy far-reaching immunity to their impact. So should one with allies that can re-supply it by land. The great British victory at Trafalgar in 1805 did little to unseat Napoleon. Nor did the

Allies' maritime power in World War II really start making itself felt before they landed on the Continent.

Likewise, naval raids on enemy territory, by acting as diversions, compelling the enemy to divide his forces, and in general acting as a thorn in the enemy's side, can bring excellent results. The Allied raid at Inchon in 1950 developed into a spectacular success. But such operations, especially if they are not supported from the land, may become prolonged and consume more resources than they are worth. The 1854 Anglo-French landing at Sebastopol ultimately ended in victory. But the 1915 one at Gallipoli did not.

Full-scale, opposed, amphibious invasions represent the crowning achievement of naval warfare. Normally only they can move war from the sea, where the enemy is not, to the land, where he is. Some invasions, such as those mounted by William the Conqueror in 1066 and by the Allies in 1942–5, were successful. Others, such as the Athenian expedition to Syracuse in 415 BC and the Japanese attempt to capture Port Moresby and Midway in 1942, ended in disaster—some more so, some less.

As many would-be invaders have found out, of all military operations amphibious ones are the most difficult. Technical problems apart, there are two reasons for this. First, the best place for disembarking the troops may not be the most suitable for continued operations on land, and the other way around. Second, no other operation is so vulnerable to counterattack both by sea and by land. Trying them without first assuring at least local command of the sea is foolhardy. But even if this condition is met, a powerful modern country with a good ground transportation network ought to be able to defeat them. So great are the difficulties that some observers now consider amphibious operations obsolete.

All this is captured in the popular pre-1914 image of the leviathan and the bear. The leviathan stood for Britain, the bear for Russia. At stake was control over Persia (Iran) and India (which at that time included modern Pakistan). Or so the authorities in London, watching Russian advances in Central Asia, believed. Yet it was hard to see how the two sides could really get to grips with each other. The leviathan, as long as it dominated the sea, was immune against either invasion or strangulation. The bear might have its overseas trade cut and suffer accordingly. However, as long as it dominated the land it could only be defeated by a massive amphibious invasion followed by extensive ground operations. Partly as a result, no war between the two ever materialized. Had it broken out, then the advantages would by no means have been all on one side.

3. Leviathans, Bears, and Birds

For millennia on end, ships used to fight ships and armies, armies. Ships could sometimes help defend coastal cities. At Acre in 1798 the British Navy, cruising near the coast, saved the city from falling to Napoleon. They could also bring up supplies, as during the Peninsular War of 1807–13; mount raids; and stage amphibious landings. But here their ability to influence land operations ended. Conversely, except for bombarding and capturing ports, straits, and the like, armies could do little to influence war at sea. Until, that is, the beginning of the twentieth century when a third species entered the fray—birds. The impact of air power on war as a whole will be explored in the next chapter. Here we shall briefly address the way it influenced war at sea.

The first to experiment with the use of aircraft—land-based ones— at sea were the Italians during their 1911–12 war against the Ottoman Empire. Though the equipment was primitive, the results were encouraging. The Italians even used their ships to refill dirigibles with hydrogen. In 1913 Greek aircraft dropped bombs on Turkish ships. During World War I all the main belligerents used aircraft and tethered balloons to look for targets at sea. Finding one, they either dropped bombs or torpedoes to attack it themselves or else used radio to direct naval forces towards it.

Both surface vessels and submarines, the latter as they surfaced in order to recharge their batteries, could be and were attacked in this way. The war also witnessed the use of the first hydroplanes and flying boats. Yet the limited range of aircraft did not allow them to operate over the high seas. To deal with this problem, some countries started putting aircraft aboard ship or, in one case, a submarine. Seaborne aircraft could and did find, reach, and attack targets both at sea and on land. However, their load-carrying capacity was limited. With few exceptions, their attacks amounted to mere pinpricks.

The first custom-built aircraft carriers went to sea in the early 1920s. The debate as to which was to be the vessel of the future, carriers or the older, gun-carrying battleships, went on throughout the interwar period. Another question that kept ministries and staffs busy was the precise relationship between the various arms and services. Land and sea forces had long been separate, a fact that *inter alia* helps explain why so few theoreticians ever took a serious interest in both. Now new problems made their appearance and demanded a solution. Landlubbers pulled in one direction, navalists in another, and airmen in a third.

Originally armies and navies, acting separately, each set up their own air arms. In the U.S. and Japan they remained in control of those arms right through World War II. The outcome was that, in effect, these countries had not one air force but two. Most other countries took a different course. From 1918 on, with Britain in the lead, they set up independent air services. No sooner had those services been established than their commanders tried to deprive armies of all but a handful of aircraft to be used for liaison, battlefield reconnaissance, and so on. Generally, with success.

Some navies also lost their air arms. The most important one was the Italian one. That was one reason why, in 1940–3, it became a model of ineffectiveness. On the whole, though, navies proved a tougher nut for air-power enthusiasts to crack than armies. The British Navy in particular fought back. In the end it succeeded in retaining its air arm. So, though its importance was minor by comparison, did the Soviet Navy. Not having carriers, it relied almost entirely on land-based aircraft and was the only service of its kind to operate long-range heavy bombers. Endless talk about "jointness" notwithstanding, the three-sided struggle as to who should be in charge of what shows no sign of ending. Fueled by different strategic visions and billions of dollars, there is little likelihood that it will.

World War II put the various kinds of vessels, as well as the aircraft that now supported and/or accompanied them, to the test. In northwestern Europe, in the Mediterranean, and in the Far East some things quickly became clear. First, surface fleets that did not have air support, whether land or sea based, were at such a disadvantage against those who had it that operating them almost amounted to suicide. Second, submarines too proved vulnerable to air power. Not as much as surface fleets, perhaps, but enough to turn aircraft and, later, helicopters into the main instrument for hunting them at sea. Third, by joining air power with submarines, especially for reconnaissance, target-acquisition, and commerce raiding, the latter's effectiveness could be enormously increased.

Before 1914 the only ways in which navies could project power were by mounting raids and staging amphibious landings. Naval air power increased this capability many times over. But for its carriers, the U.S. Navy's campaign in the Pacific would have been impossible. Yet that was but one side of the coin. Just as sea-based air power could be used against targets on land, so land-based air power could be used against targets at sea. The more so because, aircraft per aircraft, deploying them on land was much simpler and cheaper than doing the same at sea. The aircraft themselves could be larger, heavier, and capable of carrying more fuel and more ordnance.

Aircraft were also used to sow mines. As the range of aircraft grew, increasingly fleets were compelled to avoid enemy-held shallow waters and straits and limit their operations to the high seas. The introduction of reconnaissance satellites and cruise missiles from 1960 on reinforced the trend. The basic relationship between leviathans, bears, and birds remained unchanged. Nevertheless, year by year and technological advance by technological advance, the regions of the sea in which surface fleets can operate in relative safety are shrinking.

One cardinal development remains to be mentioned: the use of naval vessels, both surface and submarine, for launching ballistic and cruise missiles at inland targets far from the coast. The missiles, what is more, can be provided with nuclear warheads. Nuclear war itself will be discussed in the next chapter but one. It is, however, essential to keep in mind that, of all known ways to make nuclear delivery vehicles invulnerable and maintain a second-strike capability, putting them aboard submarines is one of the best. In part, this is because the oceans make up about 70 percent of the earth's surface. In part, because submarines remain relatively hard to detect and sink. In the future more countries are likely to acquire them. Thus the youngest arm of the fleet is well on its way to becoming the most important one.

VIII

Air War, Space War, and Cyberwar

1. Air War

A s the phrases "command of the air" and "command of space" imply, air and space war bear some important similarities to war at sea. In all three environments the most important, and often the first, thing one can and should do is to wrest operational freedom while denying it to the opponent. If not everywhere and once and for all, at least at the places and times that matter for the immediate purpose. Other similarities abound. Air and space are even more hostile to human life than the sea is. As a result, technology plays a paramount role in them.

The important role played by technology means that pilots, air crews, and ground personnel must be highly skilled. So must those who maintain and launch missiles, operate drones, and control all sorts of space assets. All this demands the kind of training only long-time professionals can receive; neither air nor space war allow room for irregulars or amateurs. Many highly skilled people will find it easier to move into the civilian market, perhaps receiving higher salaries while working under fewer restrictions. That in turn has implications for the kind of leadership they need as well as the kind of discipline to which they can and must be subjected.

Pilots have always put their lives at risk, often more so than any other warriors except infantrymen. In 1914–18 the life expectancy of young British pilots was measured in weeks. As one of them wrote:[1]

The Young Aviator lay dying
And as in the hangar he lay, he lay,
To the mechanics who round him were standing
These last parting words he did say.
Take the cylinders out of my kidneys
The connecting rod out of my brain, my brain,
The cam box from under my backbone
And assemble the engine again.
Then go ye and get me school bus
And bury me out on the Plain, the Plain.

Not so other members of the air and space community, be they mechanics, or ground controllers, or logisticians, or weathermen. Working far from the battlefield, their exposure to the *Strapazen* of war, let alone physical danger, is much more limited, often all but nonexistent. They are technicians first and warriors second. The fact that their work is unheroic does not necessarily mean it is easy. Indeed the frequent occurrence of stress syndromes among drone operators seems to indicate that war has not yet been turned into a videogame. It remains what Thucydides said it was—a human thing.

Given their limited endurance, aircraft are even more dependent on bases than ships. An aircraft running out of fuel will fall out of the sky along with the pilot and any other crew members. The same applies to unmanned aircraft or drones. Satellites are another matter again. Many have solar panels to supply their energy needs. Once in orbit, they can stay there for extended periods. Another important resemblance is that aircraft and spacecraft, as long as they remain on the ground or close to it during takeoff and landing, are vulnerable and require protection. Once they leave base, though, they are even more free than ships to move in all directions regardless of geography, topography, and manmade obstacles of any kind.

Attempts to use kites and balloons in war go back all the way to 3rd-century AD China. The first powered aircraft made their appearance during the 1911–2 Italo-Turkish War. By the end of World War I they were performing many different missions. Among them were reconnaissance, artillery spotting, liaison, air-to-air combat, and air-to-ground operations such as strafing and bombing, including such as was directed against airfields in order to gain command of the air. During the interwar period air transport, also used for casualty evacuation, was added. So was air assault, in the form of gliders and paratroopers. By 1939 air power had become vitally important; so much so that few large-scale operations, defensive or offensive, stood much of a chance of succeeding unless they were supported from the air. That still remains the case.

Early on it was thought that the speed of air power, its range, its ability to overfly any kind of terrain, and its flexibility in focusing now against one target, now against another made it an inherently offensive weapon. One which, once command of the air had been obtained, nothing and nobody, least of all civilians, could withstand. Starting with H. G. Wells' 1907 novel, *The War in the Air*, writers in and out of uniform competed in describing the horrors to come.[2] All agreed that there would be vast destruction as well as many dead and injured. However, their number would be dwarfed by vast crowds, maddened by fear, stampeding about, causing government to fail, and bringing organized life to a halt.

These expectations turned out to be exaggerated. In World War I most "strategic" attacks on cities amounted to mere pinpricks. In World War II things were different. At first, cities such as Warsaw and Rotterdam were hit so hard as to make the Polish and Dutch governments capitulate. Later, things changed. Both in Europe and Japan, massive and sustained air attacks killed hundreds of thousands of people and wrought tremendous destruction. Yet many industrial targets turned out to be harder to hit and easier to repair than had been thought. More important still, in most cases civilian morale, supported by effective civil defense, did not break down to the point where cities, let alone entire countries, ceased to function. The events of 1950–3 taught a similar lesson. Though the U.S. Air Force destroyed every North Korean city several times over, these bombings did not have a major impact on the war as a whole.

Another reason why air bombardment did not prove as effective as some had hoped and others had feared was because new weapons and equipment such as fighter aircraft, anti-aircraft artillery, and, above all, radar changed the equation. As the Battle of Britain in particular made clear, air power, far from being an exclusively offensive instrument, could be used equally well on the defense. As the air battle developed, the usual swing effect set in. Now one side, now the other gained the upper hand. In the end the "strategic" bombardment of industrial and demographic targets—read, cities—in Germany and Japan succeeded in reducing large parts of those countries to rubble, thereby making a substantial contribution to victory. However, doing so proved far harder, took much longer, and occasioned far more losses than anybody had anticipated.

Another element of air power that proved problematic was air assault. During World War II both sides used it on a considerable scale, but only at the cost of staggering losses. That was true both when it succeeded, as when the Germans occupied Crete in 1941, and when it failed, as when the Allies descended on Arnhem in 1944. The French failure at Dien Bien Phu in 1954 sealed its fate once and for all. Since then helicopters have taken over some of the roles of gliders and paratroopers. However, the range of helicopters is more limited than that of aircraft. The number of troops and the size of the loads they can carry is less, and they are far more vulnerable to enemy fire. These factors limited their use in air assault to small-scale operations. What helicopters can do well is liaise, serve as flying command posts, transport troops and equipment to any point they are needed, act as gunships, and cooperate with ground forces in general. Hence their spread added fuel to the old turf battles as to which service should control what.

Meanwhile the advent of jet engines enabled fixed-wing aircraft of all kinds to grow larger, faster, and more capable. From about 1960 on air-to-air and air-to-ground missiles started replacing cannons and bombs respectively, dramatically increasing range and precision. During the 1990s the first stealth aircraft emerged. The battle between offensive and defensive warfare, aircraft and anti-aircraft defenses, continued. The more so because roughly the same cannons, guided missiles, radar, and other electronic gear that aircraft used could also be used for shooting these same aircraft out of the sky. In so far as air offensives against countries such as Iraq (1991 and 2003) and Serbia (1999) succeeded, this was mainly because they were mounted by a superpower against third-rate military powers. In 1999 some twenty years had passed since the Serbs had received a single modern anti-aircraft missile. The Iraqis in 2003 were in a worse situation still.

What applied to "strategic" bombing and to air assault also applied to other ways of using air power. Flying low over a well-entrenched enemy capable of firing back has always been, and remains, hazardous. Overall, the best way to employ air power in conventional war may be to hit the enemy's lines of communication. The Luftwaffe in 1939–41, the Royal Air Force in the Western Desert from the end of 1942 on, the U.S. Army's IXth Tactical Air Force in Europe in 1944–5, and the Israelis during the 1967 June War all provide excellent examples of what can be done. Air power also played an important role in the 1950–3 Korean War, the 1956 Suez Campaign, the 1965 and 1971 Indo-Pakistani Wars, the October 1973 Arab–Israeli War, and the 1980–8 Iran–Iraq War.

Still, on none of these occasions did it produce decisive results. In 1973, indeed, it was the Israeli ground forces that destroyed the Egyptian air defenses and opened the way to the air force, not the other way around. If reports coming out of Iraq may be believed, Saddam Hussein during the last desperate days of the First Gulf War kept asking whether the Allies would advance on Baghdad, not whether they would keep bombing his country. The prevalence in many places around the world of intrastate wars waged by terrorists, guerrillas, and other kinds of insurgents who use the civilian population for cover has made the use of conventional air power more problematic still.

During the early 1960s, ballistic missiles started replacing manned bombers for nuclear strategic missions. Satellites did the same in respect to surveillance and reconnaissance. Cruise missiles followed in the 1970s, drones around 1980, and so-called killer drones twenty years later. Compared to manned aircraft, these machines have considerable advantages. Chief among them are the much smaller and less vulnerable infrastructures many require, dispensability, and, in the case of drones, the ability to loiter above the battlefield, as well as smaller training and operating costs. That is why the number of drones in particular has been skyrocketing. Joining the other devices just mentioned, they now threaten the very survival of manned combat aircraft. Conversely, in most of the world the extraordinary cost of such aircraft is reducing their number by about one-third every fifteen years. Even when the air forces that operate them have not seen combat.

Clearly humans, having turned the air into one more environment in which to fight, are not going to leave it again. Air power alone enables control of the air to be won. That done, aircraft can engage in intelligence, surveillance, and reconnaissance; strike ground targets; and provide certain kinds of mobility that no other service can match. In one form or another air power will remain critically important, and occasionally decisive, in conventional war. The problem is that, early in the twenty-first century, the vast majority of wars are no longer of the conventional type. Instead they consist of terrorism/counterterrorism, guerrilla/counterguerrilla, or insurgency/counterinsurgency. And in such wars, as countless cases from Vietnam to Afghanistan have shown, the contribution air power can make is much more limited.

2. Space War

In some ways space war is a straightforward continuation of air war—or so air forces everywhere claim. However, the technical problems it presents are, in many respects, even more formidable. First, so hostile is the environment that practically everything must be unmanned and will probably remain so for a long time. That implies, among other things, that damaged equipment can only be repaired with great difficulty, if at all. Second, the use of conventional fuels is strictly limited. Other energy sources such as batteries, sunlight, and small nuclear reactors are available. But each in its own way is problematic.

Third, precautions must be taken against hazards such as meteorites and all kind of debris that, moving at great speed, can easily wreck a satellite. Fourth, distances and speeds are enormous. Extraordinary precision is required. Or else even lasers, operating at the speed of light, will fail to hit their targets.

To make war in space one must get there first. The earliest ballistic missiles were built in Germany during World War II. However, they merely passed through space on their way to target without making any use of it. That remained the case until 1957 when the Soviet Union and the United States started sending satellites into orbit. Famously, all the first satellite could do was say, "beep-beep." But that soon changed. After the first reconnaissance satellites became operational during the early 1960s, the number and variety of missions they performed expanded enormously. Among them were weather prediction; communication; mapping; navigation; early warning against attack; many kinds of intelligence (including visual, infra-red, radar, and electronic); post-action damage assessment; and much more.

As of the beginning of the twenty-first century there are some international treaties in existence that prohibit various weapons, including nuclear ones, from being stationed in space. Should any country see an important military advantage in doing so, though, surely the treaties will be abrogated. During the 1980s many plans, some of them bizarre, were proposed to use satellites for intercepting incoming ballistic missiles on their way through space to target. However, the expense would be enormous and the benefits, in terms of the protection offered to ground assets, doubtful. Some of the methods proposed demanded phenomenal amounts of electricity which are unavailable in space. Others were so indiscriminate as to put their users in almost as much danger as their intended targets. These factors go far to explain why, in the end, nothing came of it.

In principle one could also mount offensive weapons, e.g. high-speed, explosive darts, to launch precision strikes at targets on earth. Doing so would not be without some advantages. Depending on geography, warning times against missile attack could be cut. The number of directions from which such an attack could be launched would be greatly increased. Either way, the problem of mounting an effective defense against ballistic missile attack, already formidable, would be made much more difficult still.

Satellites orbit at distances of 150–22,500 miles from earth. Speeds vary between 3,335 and 17,340 miles per hour. Some satellites are maneuverable, making them even harder to intercept. Nevertheless, they are not invulnerable. The first step towards making them vulnerable would be to set up a system capable of tracking all of them, constantly and in real time. Next, one would need to devise methods for attacking them. One such method is to use anti-satellite missiles launched from the ground or from aircraft. Another is "killer" satellites capable of maneuvering in space, ramming their targets or tackling them by other means. Lasers may be used to blind them; electronic war, to interfere with their communications, put them out of action, or take them over; and an electromagnetic pulse to make them break down.

Last but not least, there are the ground control centers. Without such centers there would be no way to communicate with satellites, make them perform their missions, check on whether those missions have been carried out, and learn the results. Like any other assets, such centers can be hit by aircraft or with various kinds of missiles. They could also be neutralized by a cyber-attack. However it is done, a well-designed, well-executed offensive against a country's space assets could render the armed forces that rely on them for the above-listed missions deaf and blind. Other things being equal, the greater the dependence on them the greater the danger.

That is not the end of the story. The earliest satellites served military purposes almost exclusively. Later, civilian corporations took over and often became leaders in the field. Early in the twenty-first century an enormous number of everyday services depends on satellites. Without them, there would in many cases be no navigation, no communications, no data links, no radio or television broadcasts, no banking services, no earth observation—and so many other things that it is enough to make one's head spin.

Even most of the Pentagon's own communications network, much the largest on earth, relies on civilian assets rather than on dedicated military ones. The Global Positioning System, used not only for navigation but for guiding many kinds of missiles to their targets, is particularly vulnerable. Owing to budgetary considerations, this situation is likely to persist in the future. In theory a sufficiently well-planned, broad, powerful, and sustained space offensive could paralyze the vital communications infrastructure not only of armed forces unprepared to meet it but of entire countries. It could do so, moreover, not in years, as with old-fashioned "strategic" bombing, nor in weeks or days, as with precision-guided strikes on command centers and so on, but in a matter of hours if not minutes.

Such visions are eerily like those which, in the 1930s, described the horrors of strategic bombing. Some speak of a "space Pearl Harbor."[3] They demand that resources be allocated, and measures implemented, to minimize the danger. Some countries have moved in this direction. Communications can be and are encrypted. Satellites can be hardened to make them less vulnerable to attack. They can also be miniaturized and made harder to target. Produced in some numbers, they might be stored and launched to replace those that have been neutralized or destroyed. The number of possibilities is endless. But clearly there are limits. Protecting all elements of the vast civilian infrastructure circling the earth would be comparable to rendering a large city capable of riding out a number of hydrogen bombs dropped on it. In theory it could be done; but the expense would be beyond imagination.

As of the early years of the twenty-first century, some aspects of space war are being experimented with. Others are in the process of development. Nevertheless, there is some ground for optimism. Between 1957 and 2013 no fewer than sixty-six countries launched satellites. Some used their own launchers for the purpose, but the majority employed missiles put at their disposal by others. As far as public information allows, no attempt has ever been made to shoot down an enemy satellite.

Some of the satellites are military, others civilian. Many are dual use. Military systems serve either one country only or are shared among allies. However, practically all of the countless civilian ones serve a great many simultaneously. Often that includes friends and at least potential enemies alike. Some U.S. spy satellites have been put into orbit by Russian rockets. So complex and so interwoven are the networks that many kinds of space attacks aimed at one country would undoubtedly hit others too. The result would be political complications to make the Gordian knot look simple by comparison. Neither a boomerang effect nor, perhaps, global chaos can be excluded. All these are problems that space war shares with cyberwar; and so it is to the latter that we must now turn.

3. Cyberwar

From the invention of ships millennia ago to the onset of the space age, the emergence of new technological devices has always caused war to spread into new environments. The most recent addition to the list is cyberspace, defined as the realm of computer networks where information is stored and communicated online. Some analysts expand this definition to include what used to be electronic warfare. As of the early twenty-first century it is hard to imagine any additional one into which war might spread. However, until 1990 or so space seemed to be "the final frontier." Cyberspace is an entirely man-made environment. Who knows whether others will follow it and what forms they will take?

Digital electronic computers go back to the late 1940s, networks that link them together to the 1970s. No sooner were computers linked to each other than it became clear that every network contains vulnerabilities and that those vulnerabilities could be used for intelligence purposes. The process is sometimes known as "information war."[4] In some ways, "information war" is a direct offshoot of ordinary intelligence operations, especially those that rely on electronics such as ELINT (information derived primarily from electronic signals that do not contain speech or text). Nor are the considerations governing its employment very different. In all cases it is a question of learning as much about the opponent as possible, feeding him with disinformation, and keeping information about oneself secret. The overall objective is to obtain "information superiority."

Since no physical violence is involved, information war does not amount to war any more than traditional forms of intelligence and counterintelligence, on their own, do. But at what point does espionage become war? Some kinds of computer malware can do much more than discover information stored in, or emanating from, an opponent's computers. They can also erase or modify that information, feed them with false information, cause them to malfunction, and make them break down beyond repair.

Much worse still, computers now run countless systems, both military and civilian. They range from inventories of spare parts to telephone networks; and from electricity grids to the nets that trade and banking use. Almost all these systems form part of a single gigantic Net, which means that they can be accessed through the latter. In theory, a concentrated attack on all of them, or even on many of the critical nodes that link them to each other, could quickly bring an entire country to a halt. Especially if private companies, which own most of the Net, gave the responsible government agencies of the target country access to the necessary codes.

Again, predictions and warnings in the field bear an eerie resemblance to those concerning air warfare in the 1930s. First one would prepare a detailed list of targets and assign a team of specialists to each as well as to the links between them. Many kinds of malware would have to be designed, produced, tested, and implanted in the other side's network without being noticed. At a given signal the attack would start. First military sensors, communications, command and control centers, supply systems, anti-aircraft defenses, and so on would be made to malfunction or neutralized, thus preventing them from being used in retaliation. Next, the pumps of oil refineries could be damaged, leading to fires. The turbines in power stations could be made to rotate too fast (or, perhaps, not at all), break loose from their moorings, and destroy themselves. A country's electricity grid could be disrupted. So could its health services, rail transportation system, banks, clearing houses, stock exchanges, telephone services, and so on.

With all this achieved, vital commodities such as food, water, energy, fuel, medicines, and money would quickly no longer be available at the times and places they are needed. A ripple effect would play havoc with orderly economic life and bring it to a halt. Attempting to restore it, governments would find that the bulk of their own traffic, passing through commercial channels, had also been affected. The all but certain outcome would be widespread rioting and social disintegration, to be followed, soon enough, by mass starvation. At no point would the attacker use direct physical violence: nevertheless, there could be little doubt that the polity or organization which was doing all this to another was engaged in an act of war.

Where would such an attack fit in, and how would it compare with more traditional forms of war? Presumably the greatest similarity would be in respect to the overall objective: namely, to inflict as much damage as may be needed to bend the opponent to one's will while limiting one's losses. There are, however, some problems here. Suppose a really powerful cyber-attack succeeded in creating communications chaos on the other side: would there be anyone left able to accept terms and carry them out?

As with conventional war, organization and direction are essential—individuals on their own are unlikely to cause much more than local, strictly limited, disruption. As with conventional war, uncertainty (but not friction, which is negligible) reigns. As with conventional war, intelligence, counter-intelligence, robustness, redundancy, and resilience are vital. Cyberwar also resembles conventional war in that it involves move, counter-move, counter-counter-move, and so on. In so far as any weapon can only be used once, this is even more the case in this field than in many others. Barring catastrophic damage at the outset, the long-term outcome will be, already is, a swing effect. Now one side, now the other will draw ahead.

As with conventional war, one must find a point—a so-called node, perhaps—in the enemy's array that is both vitally important and vulnerable. Some of the pairs of opposites that make up conventional strategy apply to cyberwar too. They are: maintenance of aim versus flexibility; husbanding force versus sacrificing it (since any computer used on the attack can be counterattacked, probably some computers will always have to be kept disconnected from the net, forming a reserve); the direct versus the indirect approach (as by routing attacks through third parties so as to make it harder to identify the attacker); and strength against weakness.

True, pairs whose operation depends on physical space will be irrelevant, such as concentration versus dispersion (though concentrating all one's hackers and computers at a single point may make them vulnerable to physical strike); battle versus maneuver; breakthrough versus envelopment; and advance versus retreat. All in all, though, the elements of strategy, including also surprise and the critically important interaction between the ordinary and the extraordinary, are just as relevant to cyberwar as to any other kind.

Once it has begun, cyberwar proceeds at enormous speed and often automatically. This creates a powerful push towards escalation; the more so because most decision-makers are much less familiar with the world of computing than with that of conventional war. Starting at least as far back as the Roman general Pompey, who between 66 and 63 BC conquered the entire Middle East without so much as consulting the Senate, there have been many cases when generals ignored their political masters. They picked up the ball and ran with it. Is there any reason why programmers cannot do the same?

The differences between cyberwar and more traditional forms of war are equally striking. Perhaps the greatest one is that, since practically all networks now form part of a single gigantic Net, cyber-attacks are hard to direct and to aim. Attacking one target, one may not know what others are going to be affected and how. For example, when Iran's nuclear installations came under a cyber-attack in 2010 thousands of other computers both inside and outside that country were also infected—though without suffering any harm. The "sprinkler" nature of cyber-attacks may have its advantages when it comes to gathering intelligence. It does, however, mean that care has to be taken not to hit the wrong target(s), including one's own.

Much more than conventional offensives, cyber-attacks can be prepared and launched in secret. Threat assessment is very difficult, surprise—so called "zero day" attacks—all but impossible to avoid. It is even possible that the victim will not be aware of his situation until long after the attack has started and ended. More than with any other form of war, qualitative superiority is critical whereas quantity hardly counts. Unlike their conventional counterparts, and assuming reasonable tracking occurs, offensive programs used in cyberwar can only be used once. In this sense they are more like stratagems than like weapons proper.

Systems, and there are some, which do not form part of the Net can only be attacked if they are physically accessed first. Stuxnet, the program which, with the aid of a scandisk, was literally stuck into an Iranian computer and caused much damage to its nuclear program, is a good example of this.[5] As long as there is no air gap, though, cyber-attacks can be launched from any point, in any direction, over any distance, and regardless of geographical and topographical obstacles, natural or artificial. Unlike kinetic (i.e., physical) attacks, they are also unaffected by the way attackers and targets move in space and they do not require physical movement, so they don't need transportation systems and lines of communication of any kind. Logistically their requirements are negligible. There is no need to move masses of troops and mountains of equipment. Preparing an attack may be expensive and take time. But once the weapons exist the cost of using them is also negligible. In all these ways they favor the offense against the defense, the small against the big, and the poor against the rich.

Big or small, cyber-attacks have become part of the immensely complex mosaic of modern life. Many players, each according to their objectives and capabilities, have adjusted to it and taken it in their stride. But supposing an all-out cyber-attack by one country on another, how likely is it to land a knock-out blow? The answer, based on what experience in the matter has been made public, seems to be, not very. The first cyberwar between different countries, as opposed to hacking attacks by one organization on another, took place in 2007. Russian hackers, unhappy with a decision to move a statue commemorating Soviet dead in World War II to another part of Tallinn, capital of Estonia, launched a cyber-attack on that country.

Later in the same year the Israeli Air Force mounted a cyber-attack that paralyzed Syria's anti-aircraft system, or so the media claimed. Next, it used the opening thus created to bomb and knock out the country's nuclear reactor, then under construction, without loss to itself. In 2008 Georgia came under a cyber-attack that shut down websites and blocked communications with the outside world. All this, even as Russian forces were dropping bombs and invading the country. Two years later the aforementioned Stuxnet virus, supposedly created and implanted by Israeli and American intelligence agencies, made headlines. It penetrated the computer systems controlling parts of Iran's nuclear program and alternately accelerated or slowed down the centrifuges used for enriching uranium, ultimately causing them to break down. Later two other programs, known as Duqu and Flame, were discovered. Probably they formed part of the same operation.

How much can be learnt from this limited sample is moot. In none of the attacks did those behind them admit responsibility. That did not matter much when the attack was combined with a physical one, as when Israel bombed Syria. But in the other three cases proof was hard to obtain. The hackers who attacked Estonia and Georgia may have been operating on their own. Perhaps more likely, they worked for the Russian security services which aided them and will certainly provide them with cover if anybody tries to go after them.

All the attacks, as well as others that may have taken place but remained unpublicized and perhaps undetected, took their victims by surprise. This made deterring them, preempting them, and retaliating against them all but impossible. Still, the attacks did not even come near to paralyzing the countries against which they were directed (assuming that doing so was, in fact, their objective). Though some damage was done, in all cases it was repaired with relative ease. If the Syrian reactor was an exception, then that was only because it was physically bombed.

When a 2014 newspaper headline said a bug could "take out" an entire cruise ship it was referring to a pathogen, not a deliberately induced computer failure. Since offensive cyberwarriors may not be identifiable, their operations may be less like large-scale conventional war and more like terrorism, sabotage, or crime. Clausewitz's dictum that surprise is effective in small matters but much less likely to decide big ones also seems to hold true here.[6] All this makes cyberwar suitable for waging attrition campaigns as well as salami-type attacks. Another possibility is to combine it with more conventional war, using it as a can-opener in much the same way as special forces are often used.

Many governments have set up agencies to defend their countries against cyber-attacks. They do research, provide information, sound alarms, probe the defenses for weak spots, and mandate or recommend any number of security measures. However, there are problems. The government may be reluctant to share its cyberwarfare expertise with private firms. In turn, private firms may not want to have the government snooping on them. That is one reason why some corporations would like nothing better than to break loose from the states in which they are based and set out on their own. If standards are too low, they will be useless. If they are too high, many companies will be reluctant or unable to adopt them. If they are uniform they may be dangerous; a field used to raise a single crop such as grain or corn is much more vulnerable to diseases and pests than one on which many different plants grow. Everything considered, probably the best course is to mandate minimum standards and allow each organization to work out the details.

Other things being equal, the more networked a country, nation, or organization is, the more vulnerable it is to cyberwarfare. Conversely, the best defense would be to do without computers or, at any rate, the nets that link them to each other. The latter solution wouldn't solve all problems; computers need operators, and operators can be misled or corrupted. Still, it would make the country that adopted it much less vulnerable. The former option would make it immune. But only at the cost of sending the country in question back to something like the Stone Age.

As the 2013 Snowden revelations concerning America's National Security Agency made clear, information warfare and cyberwarfare now proceed not second by second but nanosecond by nanosecond.[7] Dozens of states play the game, both offensively and defensively. So do many non-state organizations and groups. Both the damage cyber-attacks do and the cost of defending against them is in the tens of billions, perhaps more. Many pockets of disruption, some small, others large, have been created. Still, apparently no electricity grid has collapsed. Nor did aircraft collide when a computer virus para-lyzed air traffic control systems. The 2014 Russian takeover of the Crimea was accompanied by intensive cyberwarfare that involved Russia, the Ukraine, and NATO. Yet it was not hackers but troops and their weapons who prevailed.

So, using air bombardment as an analogy, unless the guard is relaxed and grossly mistaken policies are followed, a massive cyber-attack capable of dealing an instant death blow to entire countries appears rather unlikely. Nor, for all the differences between cyberwar and more traditional forms of armed conflict, can cyberwar render military history irrelevant and totally change the character of war. To that extent, it resembles all its predecessors. All, that is, except for one.

IX

Nuclear War

1. The Absolute Weapon

Without nuclear weapons, no nuclear strategy. The first atomic bomb developed power equal to that of 14,000 tons of high explosive. Some 80,000 people were killed on the spot. Later another 10,000–50,000 died of the effects of radiation. That was just the beginning. The largest bomb ever was the Russian *Tsar Bomba*, imperial bomb, of 1961. It turned out to be 4,000 times as powerful as the one that demolished Hiroshima. Since such a monster can only make the rubble bounce, nothing like it has ever entered the arsenals of the world. A number of smaller bombs, properly targeted, would be much more useful.

Just one hydrogen bomb of the right kind, detonating at the right altitude and sending out an electromagnetic pulse, could wreck countless sensitive electronic devices below, inflicting much greater damage than any cyber-offensive. It might even paralyze an entire continent. Other nuclear weapons can finish off all human—and not just human—life within a large radius while leaving structures and machines intact. Others still produce radiation that can make whole countries uninhabitable for years, even forever if that is the designers' purpose.

Our ancestors have long dreamt of devising an "absolute" weapon. To wit, one so powerful that it can finish off the opponent in a single blow, leaving him prostrate without allowing him time to defend himself or escape. Jehovah, raining down "large rocks" on the

Canaanites at Beth Horon near Jerusalem and killing more men than had perished by the sword, used such a weapon. So did Zeus hurling his thunderbolts from the inaccessible heights of Mount Olympus. So why not the sons of man as well? Now that just such a weapon exists, people have soon come to discover—in some cases much to their disappointment, no doubt—that it could not be used. And that therefore, no military strategy, at least in the sense that this term has been employed throughout this volume, was either needed or even any longer possible.

For millennia before 1945 the power of weapons kept on increasing, now slowly, now by leaps and bounds. So did their speed, range, rate of fire, accuracy, defensive strength in resisting penetration, and many other things. Especially from the time of the industrial revolution on, often a single device could wreak death and destruction of a kind, and on a scale, that used to require a hundred or more. As happened, for example, when a machine gun could produce as much firepower as a company of infantrymen and also when submarines, aircraft, space-craft, and computers took warfare into environments that were entirely new.

Each time new weapons and equipment were introduced, new doctrines, new training methods, and new forms of organization had to be devised. Things also worked the other way around. Here and there the impact went far beyond the military. As, for instance, when the invention of artillery, which only the greatest lords could afford, helped end the Middle Ages and usher in the modern age. Still technology, including chemical and biological weapons, only affected the *way* war was fought. It did not change its nature as an instrument of policy, however defined.

Nuclear weapons went much further. That was because, unlike other weapons of mass destruction, in theory and perhaps in practice they can put an end to human life on this planet. The reason why they can put an end to human life on this planet is because, unlike other weapons of mass destruction (and probably weapons of mass disruption, as cyberweapons are sometimes called), their impact is instantaneous. The victims have no time to react. Hence nuclear weapons have affected not just how war can be fought but also the reasons why it is fought and what it might be fought for. In other words, the manner in which war can be used, or even whether it can be used at all, as such an instrument.

Nuclear weapons have cut the link between victory and survival. Throughout history the victor, however great his losses, could expect to live another day. Following his battle against the Romans in 279 BC, King Pyrrhus of Epirus is supposed to have said that "Another such victory, and we are lost."[1] Had he gained his "victory" by using nuclear weapons against a Rome provided with a nuclear second-strike capability, then as surely as night falls he *would* have been lost. Presumably Roman retaliation would have been sufficiently awesome to make the war not worth fighting at all—always assuming that Pyrrhus and/or any of his subjects had been left alive and able to consider the matter.

So great was the revolution in warfare that nuclear weapons wrought, and so monstrous the possibilities they opened, that most people took a long time to comprehend the new situation. Some still don't. The critical fact of nuclear life is that it is here to stay. If the bombs are dismantled, then the reactors and plutonium separation plants, which are also used for civilian purposes, will stay. If the reactors and separation plants are closed, then the nuclear fuel already produced will remain and be all but impossible to get rid of.

Suppose we launch the fuel towards the sun, where it will be burnt. Even then the know-how needed to produce more will still be available. If all the scientists and technicians are killed, as in Walter Miller's 1960 novel *A Canticle for Leibowitz*,[2] then surely new ones will take their place almost as soon as governments, promising payment, start calling on them. In short, no disarmament program, however well intentioned, comprehensive, and successful in achieving its aims it may be, is going to eliminate the ability to rebuild the weapons. And to do so, what is more, fairly quickly.

Second, no defense or shield, however advanced, is going to render their delivery vehicles "impotent and obsolete," as U.S. President

Reagan, in his 1983 "Star Wars" speech, hoped they would become.[3] As with any other technology used in war, there is a swing effect. Now one side introduces new devices and draws ahead, now the other. All the more so because many of the same technologies used to intercept delivery vehicles, i.e. missiles, electronics, and computers, can, with some modifications, be used equally well to deliver the weapons and guide them to their targets. There are too many ways such systems can be counterattacked, swamped, and spoofed. Often not even the best intelligence can tell when a "window of opportunity" is created, just what it consists of, how long it lasts, and when it is closed.

At a deeper level still, the problem is not technological but logical. Not even a zillion successful anti-ballistic missile tests can *guarantee* that the zillionth and first will also be successful. For most purposes this kind of uncertainty makes no difference. Either because we have no say in the matter, as with whether the sun will rise tomorrow, or because we can more or less live with the consequences. Not so in the case of nuclear weapons. A decision-maker, told that he can rely on his country's defenses and safely launch an attack, should always ask his interlocutor how he can be *certain* the defenses will work as planned. Given the risk of having one's country turned into a radioactive desert, and absent a guarantee much stronger than any "ironclad" one can be, it is very hard to see what purpose launching a nuclear war could serve.

For a non-nuclear state to launch a serious attack on the vitals of a nuclear one would be madness. Assuming a reliable second-strike capability—which, however, is by no means easy to achieve—for a nuclear state to do the same on another nuclear state would be greater madness still. That is not to say that defensive systems are necessarily useless. Far from such being the case, they may reinforce deterrence by increasing the attackers' uncertainty as to whether his missiles will, in fact, knock out the opponent's second-strike capability and avoid retaliation.

None of this means that a nuclear attack by one country on another, launched by some mad dictator prepared to risk retaliation, is inconceivable. Had nuclear weapons really been unusable, then they could not have deterred anybody and anything. Presumably leaders and countries would have gone on fighting each other on a growing scale, and with growing ferocity; just as they had always done before Hiroshima and Nagasaki put their noses to the wall.

Do the facts of nuclear life seem convoluted? That is because they are. The argument as to just how nuclear weapons can be used, if they can be used, has been going on for a long time. Until it is resolved, if it is resolved, all we have to go on is the plain fact that, though the number of nuclear countries has been growing, no nuclear war has broken out. Apparently what has prevented it is the fear of escalation. To wit, the clear possibility, likelihood even, that if one nuclear weapon is used by anyone, then they will all end up being used by everyone else who has them. Perhaps without granting humanity a decent interval to settle its affairs.

By a rough estimate, the total number of nuclear devices in the arsenals of various countries early in the twenty-first century is about 15,000.[4] That is a terrifying prospect indeed. To use the Cold War-vintage phrase, the outcome of a nuclear confrontation could be "the end of civilization as we know it." Tens, perhaps hundreds, of millions would be killed outright. Dust clouds could block out the sun. Global cooling could lead to crop failures, causing social life to disintegrate as people fought each other to the death before they all starved. The survivors would die of radiation. Including not just that released by the bombs themselves but, perhaps, much-increased ultraviolet radiation caused by the destruction of the ozone layer as well.

The few survivors would be powerless in the face of wind, water, snow, ice, heat and cold, freezing, melting, rust, earthquakes, and tsunamis. Over time, these things would destroy most of our buildings, roads, and machines. Most large animals would also die. In the end, the only forms of life left on our blue planet might well be insects, grasses, and bacteria. As Clausewitz says, walls exist primarily in people's minds. Once they have been demolished, putting them back again is all but impossible.

2. Welcome, Dr. Strangelove

The nuclear stalemate that established itself from the early 1950s on did not satisfy everybody. After all, if the weapons were unusable, why spend tens of billions on them? In particular the U.S., as the most powerful nuclear country, always worried about being outnumbered by the Soviet Union's conventional forces. It consistently refused to provide a no-first-use guarantee as several other countries, including most recently North Korea, did. Instead it started looking for a way to use nuclear weapons as if they were non-nuclear; in other words, without excessive risk of escalation.

Finding ways to make the world safe for nuclear war was the task of entire regiments of Dr. Strangeloves.[5] Some wore uniforms, others, like the character in the film of the same name, business suits. Their first priority was to promote smaller, more agile nuclear weapons. So small that one could "credibly" threaten to use them—and, if necessary, turn the threat into reality.

Some of the weapons could be fired from huge artillery barrels. Others could be dropped or launched from fighter-bombers, others still delivered by short-range ballistic missiles. One, said to have had a yield of just 10–20 tons of high explosive, could be launched on the battlefield by three soldiers in a jeep. But what if the soldiers, shaken and perhaps left incommunicado in the confusion of battle, launched the weapon without authorization and/or at the wrong target? And what if they and their jeep were captured? In the West, by the second half of the 1960s enthusiasm for tactical nukes, as they were known, was on the wane.

Towards 1960, some proposed agreements that would limit the size of the weapons as well as the targets against which they could be directed. It might, for example, be possible to negotiate a pact for banning devices with a yield greater than 100,000, or 500,000, tons of TNT. Other treaties might establish nuclear-free zones in which armies and navies could fight each other to their hearts' content. Others still would establish lists of targets, such as cities, that should be left alone. However, powers behaving like scorpions in a bottle could hardly reach such deals. Nor did they.

What bilateral agreements were put in place, such as the 1972 Strategic Arms Limitation Treaty, did little to solve the problem. In their absence, unilateral measures to limit nuclear war were proposed and, to some extent, adopted. Among the earliest was counterforce. Under this doctrine an American strike would be directed not against cities, such as Moscow or Leningrad, but solely against nuclear forces and bases as well as other military targets. All in the hope that the other side would understand the "signal" and spare Washington D.C. and New York. If he did not, then enough missiles and warheads would be kept in reserve to do what had to be done.

Another option, much discussed during the 1960s, was "flexible (or graduated) response." Instead of using overwhelming force ("massive retaliation") to counter any attack, any such attack would be met by a counterattack tailored for the purpose. Hopefully this would limit the war to the battlefield and the zone of communications behind it; sparing the enemy's homeland and encouraging him to do the same. The advent, from the 1970s on, of cruise and ballistic missiles sufficiently accurate to target individual buildings brought to the fore proposals for "decapitation," i.e. striking the enemy leadership so as to leave him headless. There was also talk of "nuclear shots across the bow," a term that requires no explanation.

Though much less is known about them, other countries may have adopted similar doctrines and shaped their nuclear forces accordingly. Fortunately for the world, not one of these proposals was ever put to the test. Nor is this an accident. Throughout history, the introduction of new weapons has shocked those exposed to them. The French at Crécy in 1346 were astonished by the English longbows which decimated the flower of their knighthood. The Germans in 1943 were thrown into confusion by the Allied anti-radar measure known as "Windows." It was Windows which enabled the destruction of Hamburg, leaving tens of thousands dead. Yet seldom did the shock last for long. Either the surprised side would copy the weapon in question, or he would develop countermeasures. In both cases the effect was to make fear diminish and wane.

To date, the only exception to this rule are nuclear weapons. At first, with experience lacking, the way those who owned the devices handled them was incredibly, terrifyingly lackadaisical. Perhaps only a miracle saved the world from an accidental explosion. Even worse, throughout the 1950s exercises were held to test the doctrines mentioned above. Some armies detonated weapons, then made troops drive through "ground zero" to prove that doing so was risk-free. Next they went on to reorganize their forces to enable them to survive and fight in a radioactive environment. Nor were preparations for nuclear war limited to the military. In a number of countries "atomic" shelters were built and civil defense drills held in which schoolchildren were taught to hide under their desks.

According to Robert Gates, Secretary of Defense under Presidents Bush and Obama, the Joint Chiefs of Staff unanimously recommended to President Eisenhower that he use nuclear weapons to save the French in Vietnam.[6] Repeatedly the superpowers engaged in "brinkmanship," another term that needs no explanation. Then came the Cuban Missile Crisis, in which U.S. President Kennedy and Soviet leader Khrushchev took the leading roles. It ended with a whimper rather than a bang; but not before ten absolutely terrifying days had passed. At one point only one officer out of three aboard a trapped Soviet submarine stood between the world and a disaster beyond any in human history. Certainly in the two countries involved, and probably in others as well, the experience appears to have brought about a lasting mental change.

Since then the number of nuclear-armed countries has grown from four to nine. At least one of these is a horrible dictatorship. Two others are not far behind; one, Pakistan, is chronically unstable. Nevertheless, explicit threats to use them, let alone use them offensively in a first strike, have become rare. A few 1962-style nuclear confrontations in which the two sides looked eyeball to eyeball have taken place. But at any rate they did not at least involve the largest nuclear powers of all.

Nor was the change simply rhetorical. Behind a heavy veil of secrecy, procedures for handling the weapons were tightened. An important step in this direction was the introduction of so-called Permissive Action Links.[7] Their purpose was to ensure that the devices would not explode even if they were accidentally dropped or launched; and even if they were hijacked either by their operators or by others. Others included agreements on the establishment of nuclear-free zones near the frontiers; mutual promises not to attack nuclear installations in case of war; and advance mutual notification of imminent tests both of the delivery vehicles and of the bombs themselves. To enable the sides to communicate in an emergency, "hot lines" were set up and regularly tested.

Expanding like ink stains, all these measures are mere palliatives. Should a *really* serious crisis occur, then it is possible, indeed likely, that none of them will be of much avail. The entire purpose behind them is to ensure, as far as possible, that such a crisis should *not* occur. Either way, their introduction seems to show that fear of nuclear weapons, rather than diminishing in time as with all previous ones, has increased. Reports concerning the spread of cancer among those who received radiation during the nuclear tests of the 1950s helped fan these fears. So did accidents such as those at Chernobyl in 1986 and Fukushima in 2011, when damaged reactors spewed or leaked radioactivity, making their surroundings uninhabitable. In many countries fear helped give birth to nuclear disarmament movements. That in turn forced some leaders to pay at least lip service to the matter.

In the end, fear of what nuclear war might do brought stalemate. As more countries acquired the weapons, fear spread from the Superpowers outward. Year by year some strategists around the world drew attention as to how "delicate" the balance between any two or more nuclear countries was and how easily it might be upset. So far, though, fear has proved powerful enough to prevent the weapons from being used.

Fear of escalation also limited the nuclear powers' willingness to wage conventional war against each other. This reluctance did not affect all the armed services to the same extent. The first to feel the impact were the land armies of contiguous nuclear countries. They could no longer seriously fire at each other, let alone advance far into each other's territory. The largest incident of this kind, the so-called Kargil War of 1999, only involved between 1,000 (say the Pakistanis) and 5,000 (say the Indians) troops.[8] Certainly the Pakistanis only advanced a few miles into the Indian subcontinent. Next came navies. Given their inability to find shelter and their vulnerability, how any surface vessels in particular can expect to survive a nuclear war has long been a riddle. Still, owing to their ability to act far from home, navies, even in a nuclear world, possess a certain flexibility that armies do not have. They can do such things as showing the flag, delivering warning shots across the bow, imposing blockades (as in the Cuban Missile Crisis), and the like. In 2010 North Korea even got away with sinking a South Korean warship, the *Cheonan*.

Air forces too have retained a degree of freedom. As by overflying territory claimed by the opponent and daring him to react; as the U.S. and Japanese air forces did in connection with a number of disputed islands in the South China Sea. The same applies to space. Of the nine members of the nuclear club, only three have any kind of anti-satellite capability. In the unlikely case they go to war against the remaining six, or in case some of the latter go to war against each other, their satellites will continue to orbit. Cyberwar, which owing to its stealthy nature is very hard to deter and whose effects may only be detected late, if at all, has been affected least of all by the nuclear threat. One could, indeed, argue that, along with terrorism, these qualities make it into a method par excellence for waging war in the nuclear age.

When everything is said and done, in *no* case have two nuclear countries done more than engage in skirmishing. So far down has fear of escalation spread that even the Central Intelligence Agency and the Committee for State Security (KGB), organizations whose relations were characterized by little but hostility, agreed not to assassinate Russian and American agents respectively. So considered, as of the early twenty-first century nuclear proliferation may turn out to have been the best thing that has ever happened to humanity. Certainly it is the most important factor separating the post-1945 epoch from all previous ones—a game changer, if ever one there was.

The evident decline of major war between major powers has not taken place without cost—many would say, a very heavy cost indeed. Beginning in 1945, humanity has been living under a Damocles sword. Several decades' worth of experience and numerous precautions notwithstanding, there neither is nor can be a guarantee that the sword will not come crashing down at any moment. The price of security is insecurity. That of peace, the risk of near-instant, near-total, annihilation.

3. Deter and Compel

Both deterrence and compellence are as old as history. At times, by putting their ability to wage war on display and uttering threats, rulers have sought to avoid being attacked. At others they used similar methods to intimidate their opponents in the hope of gaining this objective or that without using force. Before 1945 nobody thought such things should be included under the rubric of war. That is why neither Sun Tzu, nor Clausewitz, nor any of their successors until that date, so much as mentions them.

In 1946 the situation changed. Impressed by the fate of Hiroshima and Nagasaki as well as the subsequent Japanese surrender, some analysts realized that nuclear weapons could hardly be put to military use in war. At the same time, other analysts started looking for other ways to use these weapons. Deterrence and compellence qualified. Both were brilliantly treated in the works of the greatest post-1945 strategic theorists: Bernard Brodie (1910–78) and Nobel Prize winner Thomas Schelling (1921–).

Since they do not involve the use of physical violence, neither deterrence nor compellence are war. To the contrary, it is when compellence and deterrence fail that war breaks out. They do, however, follow the principles of strategy. That includes the need to be as strong as possible, both quantitatively and qualitatively. It also includes challenges in the form of heavy responsibility, uncertainty, and perhaps friction among the decision-makers on each side, and the series of mirror images generated by "I think, you think..." as well as several (though not all) of the pairs of opposites we have postulated. That, as well as the role of deterrence and compellence as substitutes for nuclear war in particular, is why they have been included here.

To repeat, rulers and polities have always used threats and military displays to dissuade their enemies from going to war against them. Just as often, they did the same to make them grant their demands. At times the threats worked, at others they did not. Hitler in 1938 successfully threatened war in order to compel Czechoslovakia and its Western backers, Britain and France, to accept his demands. A year later he tried the same strategy on Poland. Failing to get his way, he unleashed World War II instead.

These cases, and many others involving compellence, are clear. Either the threatened party surrenders, or it does not. With deterrence things are more complicated. A ruler or polity who does not launch an attack on others can always claim that they had never intended to do so. In other words, that deterrence never worked on them; and that those who claimed to have used it against them had delivered a blow in the air.

So how have deterrence and compellence changed since Hiroshima? The answer is simple: the stakes are infinitely higher. Hitler, waging total war in a pre-nuclear world, said that, if the worst came to the worst, the German people would survive. He proved to be right. But had he used nuclear weapons against nuclear-armed opponents, the outcome could have been different. Would he have been deterred in the same way that dictators such as Stalin, Mao, and Korea's Kims were? He was definitely bad; but he was not mad. Nuclear weapons seem to bring even madmen to their senses. Hence the answer is that he almost certainly would. Some have the fortune, or misfortune, to be born before their time.

Deterrence and compellence are not symmetrical. The former's role in the post-1945 world has been enormous beyond measure. Starting at Kadesh in 1261 BC, when the Egyptians and the Hittites battled over control of the Middle East, great powers have never ceased waging war on each other. Judging by that fact, war, quite possibly nuclear war, between their present-day successors should have broken out long ago. As countless publications, the construction of civil and antiballistic missile defenses, and the proliferation of nuclear disarmament movements testify, many people expected it to. After the Cold War had ended some did so still. Nuclear deterrence did not prevent all wars—far from it. But it almost certainly did prevent the largest and most terrible ones of all. Hence we have a reasonable hope that it will continue to do so in the future.

To deter attack, our primate ancestors used to bare their teeth. They made their hair stand on end, emitted frightening noises, beat their chests, and waved branches as chimpanzee males sometimes do. *Homo sapiens*, thanks mainly to his mastery of language, has a larger repertoire. We can try to get our way by putting on an air of insouciance and invulnerability. We can also do the opposite and pretend to be mad (in both senses of the word). We can burn our bridges behind us. We can play a game of chicken. We can install a tripwire and tell our opponent, along with the rest of the world, that even a small attack on us may, indeed will, lead to escalation. We can explain to the opponent that, if he does X, things may escape our control. Thomas Schelling, seeking to explain how uncertainty could be manipulated to one's advantage, called this, "the threat that leaves something to chance."

For decades on end, that was how both the U.S. and the Soviet Union played the game. Where the latter differed from the former was that it never tried to "decouple" nuclear from conventional war, as the phrase went. From 1950 until 1980 or so senior Soviet commanders always insisted that any war would turn nuclear from the start. As a result, officially at any rate, they integrated their nuclear forces with the rest. As their military power peaked during the 1980s they showed some interest in making an eventual war remain at the conventional level. It did not last; the collapse of that power during the 1990s made them feel vulnerable and again increased the emphasis on tactical nuclear weapons. The zigging and zagging may not yet be over. But did it really matter? Major war, as we now know, did not break out.

In theory one could have gone further. There was occasional talk of a "doomsday machine." A doomsday machine was not simply a mechanism capable of wrecking the world. It was that, but it was also one so designed that, should an attack take place, an "automatic" response would be launched either before or after the missiles hit their targets. Technically it could be done. However, it seems that no country went ahead and built such a machine. First, the ever-present possibility of errors made it much too dangerous. Second, by taking the decision away from the leaders, it would have vastly diminished their power.

In practice, so great was the fear of nuclear war that even countries with only a handful of bombs never came under serious attack by their much more powerful enemies. Much the same applied to those strongly suspected of having them. The prime examples are Pakistan and, later, North Korea. The latter in particular has often provoked the world. Yet it has retained its immunity; unless it decides to carry out some of its frequent threats, it is likely to do so in the future too. In so far as quantitative differences matter less, deterrence in the nuclear age proved much easier than in any previous one. So great is fear of a nuclear exchange that, in war games designed to test the problem, making players press the button has proved almost impossible. Given the overriding common interest to prevent a nuclear war at practically any cost, "using" the weapons to deter nuclear enemies is relatively easy. For the same reason, doing the same to compel the opponent to do this or that is almost impossible.

Originally only one country had the bomb. That should have enabled it to lay down the law to the rest; except that, as quickly became clear, it did not. At first that was because neither the number of bombs in America's arsenal, nor their power, nor the range of the delivery vehicles that carried them, sufficed to bring a heavily armed superpower to its knees or prevent it from using its formidable conventional forces to wreak havoc. After the Soviets detonated their first nuclear weapon in 1949 the balance of terror was established. Compelling a nuclear country to submit to one's will is very hard, usually impossible.

Worse, attempts at compellence, if pushed too hard, may make the defender lose his nerve and launch a preemptive attack. The way to circumvent this problem is to adopt salami tactics. Instead of delivering a single powerful blow, the country that holds the initiative can advance piecemeal at a sub-nuclear level. It can present its opponent with an accomplished fact in a peripheral field or area and dare it to respond. If necessary it can even retreat, only to resume the struggle later. Many nuclear powers have played that game. The Soviet Union did so against the U.S. and the U.S. against the Soviet Union. China played such a game against the Soviet Union and the Soviet Union against China; Pakistan and China against India; North Korea against South Korea and the U.S.; and, increasingly, China against the U.S. and the U.S. against China.

As the Cuban Missile Crisis in particular proved, at times so great are the risks as to be almost beyond human comprehension. Yet in all cases without exception the gains, if any, were marginal. Defeat, real or perceived, in one round will drive the loser to redouble his efforts in preparation for the next one; hence they also tend to be temporary. Depending on one's point of view, the games of deterrence and compellence may be magnificent, fascinating, foolish, or shockingly inhuman. But whatever they are, they are not war.

X

War and Law

1. The Principle of the Thing (a)

As we saw, war is or should be governed by policy/politics. It is a collective activity involving two sides, sometimes more, willing and able to use almost every means in fighting each other. Its purpose is to bend the opponent to our will; its cardinal instrument is physical violence, i.e. killing people and destroying things. War differs from games, some of which are also violent, in that there is no clear definition of victory. It differs from crime, even large-scale crime such as gang warfare, in that it is socially approved, if not by the law then by at least a considerable part of the population. The last-named facts oblige us to examine the relationship between law and war in some depth.

Of our two models, Sun Tzu deals almost exclusively with how to prepare a war and win it. Consequently he focuses on its "grammar" and hardly mentions law at all. In his first chapter, though, he says that the side likely to win must enjoy the favor of heaven. Later he says that "those who excel in war first cultivate their own humanity and justice and maintain *their* [my emphasis] laws and institutions," making their governments invincible.[1] To prevent war from degenerating into a leaderless, useless, vicious, brutal free-for-all, some kind of norms must be imposed, enforced, and, above all, internalized.

Clausewitz for his part did devote a sentence or two to the problem.[2] But only to add that law scarcely weakens the elemental violence of war and that to let it govern one's conduct in war was the worst error anybody could commit. The historian Geoffrey Parker (1943–) went further. Speaking of the law of war as it existed in the early modern age, he called it "the etiquette of atrocity."

In truth there are three reasons why the role of law in war is critical. First, as has been recognized at least from the time of Cicero on, war is not merely, some would say not even mainly, an armed struggle between enemies. It is that, but it is also a legal state: either because it has been formally declared or because, as Hobbes says, its existence is "sufficiently known."[3] In war many actions that are normally prohibited, killing above all, suddenly become legal, even desirable. In theory, and often in practice as well, a state of war can exist when there are no hostilities at all, at least for a time. An excellent example is the so-called Phony War of 1939–40. For nine months two armies, the French and the German, each counting several million men, faced one another while hardly exchanging a shot. Does that mean that their countries were not at war?

Second, as Sun Tzu recognizes, the need for law stems from war's nature as an organized, collective enterprise. A one-man army needs no law. An army of any size needs it very much. There must be rules, or at least a clear consensus, as to who is permitted, and supposed, to do what to whom and why; on the orders of whom, under what circumstances, by what means, and to what purpose; briefly, as to who must be decorated and who must be hung. Other rules govern the way the legal situation changes—how war is opened and how it is brought to an end. The law of war provides answers, or tries to provide answers, to these and countless other questions. It creates a template, so to speak, as to what war is and how it ought to be waged, thus saving those who have it the need to think out everything from the beginning.

A gathering, or assembly, or whatever one may call it, which does not have such rules or does not obey them, more or less, is not an army at all. As we saw, it is a crowd or mob incapable of taking coordinated action. With people running about like headless chickens, much if not most of the time it is incapable of taking action of any kind. Such a mob cannot serve as an instrument of policy, let alone an effective one. Often it is as dangerous to those who form it as it is to anyone else.

The third reason why the law of war is indispensable is as follows. The Old Testament tells us that the Lord told the Israelites to wage eternal war on the Amalekites. The latter were a desert-dwelling people who had treacherously attacked them in the rear as they, the Israelites, were escaping from Egypt, killing the weak and the helpless. Later the Lord commanded King Saul: "Go and smite the Amalekites, and utterly destroy all that they have, and spare them not; but slay both man and woman, infant and suckling, ox, and sheep, camel and ass."[4] Only then would the war come to an end. In the eyes of the Lord, failing to bring it to such an end was a punishable offense.

That, however, was an exception. There are two reasons why war is hardly ever fought *à outrance*. Neither is altruistic or soft-hearted; both are firmly anchored in self-interest. First, most victors are sensible. Just as prisoners are promised an early release if they behave themselves in jail, so enemies can be encouraged to surrender by promising them quarter and good treatment. Doing this will save the winning side casualties and effort. Second, by taking care of the enemy wounded, for example, armies may hope for reciprocity in case fortune turns against them. One way or another, in the vast majority of cases a decision concerning the way survivors are going to be treated will have to be made and enforced. Moreover, surrender is only possible if some mutually understood method of communicating with the enemy exists. Otherwise every war, becoming "absolute," would last forever or at least until every person on the opposing side was killed.

These considerations, as well as a host of similar ones pertaining to other matters, explain why the law of war is as old as history. Judging by the way the Australian aborigine tribes, as some of the simplest societies on earth, did these things, the earliest attempt to surround war with rules was by dividing it into several different kinds. The

Murngin people of Arnhem Land, Northern Australia, had no fewer than six different forms of war, each one sufficiently distinct to be called by its own name.[5] Each was supposed to be employed for a different purpose, against a different enemy, and to be governed by a different set of customary rules. Some forms were lethal, others so much less so that they are best described as games. The rules in some cases even prescribed "victory points" to decide who had won and who had lost.

Returning to the Old Testament, we find a somewhat similar system. The most extreme form of war is the one waged by the Israelites against their hereditary enemy, the Amalekites.[6] This was however only one of the three kinds of war mentioned in the Old Testament—and the least common by far. The other two are *milhemet mitzvah*, literally "war by [divine] commandment," and *milhemet reshut*, "permitted war." Subsequent Jewish scholars, including the great Maimonides (AD 1135–1204) went further, dividing *milhemet mitzvah* into several subtypes.[7] Each type could only be proclaimed by a specific agent or body, for a different purpose, and against a different enemy. Each also had a different point at which it could be brought to an end or had to be brought to an end.

Since there was no Jewish army at any time between AD 135–7, when the last major Jewish revolt against Rome was suppressed, and 1948 when the modern state of Israel was founded, the issue became dormant. Yet no sooner did Israel arise than the laws in question became the topic of heated disputes both among rabbis and between the rabbis and the secularly minded authorities of the new state. What gives the question its importance is the fact that, unlike most other advanced countries, Israel does not recognize a distinction between state and religion. A considerable part of its population is orthodox. Of those, not a few put the authority of their rabbis over that of the state. As a result, any Israeli politician or senior commander who ignores the issue does so at his peril. Obeying one set of laws may bring him into conflict with the other, both at home and internationally.

Many other peoples also recognized different kinds of war, governed by different kinds of rules. Most codes were rooted in religion and distinguished between different enemies according to the gods they believed in. But starting with Hugo Grotius (1585–1645), secular law began to take over. Increasingly blind to creed and, much later, race and color too, it tried to lay down "universal" criteria supposedly binding on all states and peoples. The oldest laws were customary and oral. Subsequent ones were written down. Some originated in bilateral agreements that were later expanded to other countries as well. In the end many of them were put under the aegis first of the League of Nations and then of the United Nations.

Last but not least, combat can and does make people go berserk. That is what happened to Homer's greatest hero, Achilles. Early in the *Iliad* Achilles is presented as a man among men: eloquent, proud, touchy, but human. Next, fired by the death of his bosom friend Patroclus, he becomes a monster. Blind and deaf to anything but his terrible rage, he goes on a killing spree unparalleled in the whole of literature. Refusing to give quarter, his brutality is appalling. So much so that the river Scamander, running with his victims' blood, breaks its banks and tries to drown him.

Achilles' rage was soon spent. In ten years of war it cannot have lasted for more than a couple of hours, the limit set by our hormones. Soon sorrow took over. It softened his heart, making him follow custom, take a ransom, and grant his strongest enemy a decent burial. Only then did he feel sufficiently clean to follow his mother's advice: to eat, drink, and seek solace in the lap of the "soft-cheeked" woman, Briseis.[8] With her help, his re-integration into humanity is eventually completed.

Such frenzied moments apart, it is hard to find any aspect of war *not* governed by some kind of law. That is why rulers and commanders have always surrounded themselves with legal advisers and why those advisers have written countless books about the subject. Even those who refuse to recognize the law well understand its importance—or else why bother to inveigh against it?

2. *Ius ad Bellum*

As we saw, tribal people—both those of the Old Testament and some Australian aborigines—sometimes thought in terms not of a single code of law governing war as a whole but of many different ones. Each code applied to a different kind of war waged against a different people for different reasons in a different way. By contrast, the Western legal tradition has aimed at creating a single body of law applicable to all war. The most basic distinction was the one between *ius ad bellum*, the right to (wage) war, on one hand and *ius in bello*, right *in* war, on the other. The expansion of Europe from about 1500 on has spread this tradition over most of the world. That is why we shall follow it here.

Ius ad bellum addresses two cardinal questions. First comes the need to determine who does, or does not, have the right to wage war. Second, assuming somebody does have that right, one must ask what makes the war he wages a just one. Concerning the former question, the basic distinction has always been between those who ran all sorts of polities and those who were simple members of the same polities. The former had the right to declare war and wage it; the latter did not. Should they resort to force nevertheless, then again two possibilities existed. If they directed it against the rulers it was rebellion or civil war. If against each other, it was crime. Either way its use called for punishment; though punishment was normally more severe in the case of rebellion than in the case of crime.

In 1576 the French jurist Jean Bodin, in his famous *Six Books of the Republic*, popularized the term "sovereignty".[9] Rulers who owed allegiance to nobody and had no earthly superior over themselves were said to be sovereign. Those who did owe allegiance and did have such a superior were not. The first and most important right that sovereigns, unlike anyone else, possessed was to lay down the law. The second was the right to declare war and use it as an instrument of policy.

The trouble was that many aspects of feudalism were still alive. Europe was made up of an enormously complex network of different polities, some sovereign, some only half or one third or one quarter so. Each was ruled by a person or, much less often, persons, with different rights and duties. Taking their cue from the ancient Greek tyrannicides, some Protestant theologians even tried to abolish the distinction between war and rebellion. They argued that subjects, provided they were serving the True God and resisting oppression, had the right to make war on their rulers. As a result it took time and much more scholarship, notably that of Hugo Grotius, before the distinction was fully applied. Only after 1650 did Bodin's views on sovereignty come to be generally accepted—as is clear from the fact that his works ceased to be read.

At some time around 1750 Frederick II of Prussia called himself "the first servant of the state."[10] As this statement suggests, there was by now a growing tendency to transfer the attributes of sovereignty from the ruler to the polity he headed. Some polities were faster off the mark than others. However, by the middle of the nineteenth century those which had not yet made the change were widely seen as backward, despotic, or both. There things rested until 1945. Starting with the establishment of the League of Nations in 1919 and the Kellogg-Briand Pact in 1929, some attempts to limit the right of sovereign states to declare war and wage it were made. Not with much success, as subsequent events showed.

What changed the situation was World War II. At the various war crimes trials of 1945–8, some of the accused were charged with initiating and waging "aggressive" war. The precise meaning of "aggressive" has never been defined. However, the clear implication was that even sovereign states only had the right to wage "defensive" wars—and nobody ever doubted that, at the time their rulers did what they did, Germany and Japan *were* sovereign states. This principle was embodied in the United Nations' Charter and later re-affirmed several times. In response, one state after another started abolishing its war department, office, or ministry. Their place was taken by "defense" departments, offices, and ministries. The new organizations differed from their predecessors in that they included all three military services—army, navy, and air force. That apart, the change was almost purely cosmetic. War remained war, and so did the activities that took place in it.

In a different way, the change had enormous implications. For millennia on end, war had been one of the principal instruments, often almost the only instrument, by which rulers and polities exchanged territory among themselves. Now that all wars suddenly became defensive, or at any rate were declared to be so by those who started them, it followed that the use of force to conquer and annex territory could no longer be tolerated. To be sure, here and there a country still tried to do so. The best example is Israel. In 1967 it launched a preemptive war. The outcome was the occupation of the Sinai, Jerusalem, the West Bank, and the Golan Heights. In 1979–82 the Sinai was returned to Egypt. The rest remain under Israeli rule, which, however, is not recognized by a single other state.

The other change in *ius ad bellum* that took off during those years was as follows. From the time of the Swiss lawyer Emmerich Vattel (1714–67) on, it had always been assumed that occupied peoples did not have the right to rise against their occupiers. If they did so nevertheless they could be treated according to "the law of war," which in practice usually meant no law at all. In return, "military necessity" permitting, they and their property were supposed to enjoy immunity against war's worst horrors. Now that the distinction between aggressive and defensive war had become established, and annexation was prohibited, it followed that those whose land was occupied by others *did* have that right.

Even during World War II itself, many "resistance" movements in Axis-occupied countries took the law between their teeth. As a result, when those countries' own colonial subjects started using similar methods against them during the late 1940s and 1950s, they found it very hard to object. To date, few if any "liberation movements" have gained full recognition as sovereign polities. Increasingly, though, their members, instead of being treated as criminals, were given the kind of rights that ordinary criminals do not have.

That was not the end of the story. In 1990 the U.S., by asking the Security Council for permission to take the offensive against Iraq and getting it, set a precedent. More and more, the right to authorize war passed from individual states to the United Nations Security Council, albeit if the Council was unwilling, one might perhaps use something like the "Coalition of the Willing" to provide the necessary cover. Briefly, as states were losing some of their rights in going to war, all sorts of other organizations started taking a hand in the game. Either they claimed the same rights as states possessed, or they set themselves over states. In their totality, these changes are nothing short of momentous.

Intertwined with the question as to who has the right to wage war was another one, to wit: what it takes for a war to be *just*. By definition, only those who have the right to wage war can do so in a just manner. The opposite, namely that any war, provided it is waged by such a person or authority, is just, does not apply. Virgil in the *Aeneid* even seems to have two separate kinds of afterworld.[11] One for those whose fight was just; another for the rest.

Drawing on previous Roman law, Saint Augustine (AD 354–430) described war as an evil to be engaged in only out of dire necessity.[12] A just war had to grow out of man's good intentions, not his sinful nature. Last but not least, a war was just if it was used to do away with wicked persons, especially heretics who were hateful not just to their fellow humans but to God. Some nine centuries later the great scholar Saint Thomas Aquinas took up the question. First, war should be waged not for gain or "interest," as we say today, but for a good and just purpose, such as punishing evil, or correcting a wrong that could not be set right by other means. Second, since a just war must start out with some good intention, it must not be fought in so brutal a manner, and carried to such extremes, as to cancel out and over-shadow that intention. Much later, this idea came to be known as proportionality. Stripped of their original Christian dress, both arguments are still alive and well. From 1990 on, the first one in particular has been used by "the West," with the U.S. at its head, to justify a whole series of wars of intervention.

Briefly, *ius ad bellum* is supremely important. Without the distinction between those who did and did not have the right to wage war, and between just and unjust wars, society would be indistinguishable from a band of robbers and could not exist. It is true that Sun Tzu does not use this terminology. Yet he does seem to hold that a just war is one which, based on harmony between a ruler and his subjects, enjoys the favor of heaven as explained above. We may safely assume that heaven would not bless an unjust war unrightfully launched or barbarously waged. To that extent, this is as good a definition as may be had.

As noted in the introduction, Clausewitz for his part simply puts the question aside. For him, though he never says so, might is right. That view is at odds with the fact that, from the beginning of history, no person and no soldier has ever been prepared to risk his life for a cause he clearly knew, or felt, to be unjust. True, discipline and propaganda, especially propaganda that is based on religion, can do a lot to mobilize a society for war. However, discipline will come up against the fact that nothing a soldier's own commanders can do to him will be worse than that what the enemy is going to do. And those responsible for propaganda will learn that there are limits to what people will believe, how intensely they will believe it, and, above all, for how long they will do so. To that extent Clausewitz's approach, which his successors had the effrontery to call "realist," was and remains anything but.

3. *Ius in Bello*

Even supposing war is waged by appropriate authority and for a just cause, not everything is permitted in the waging of it. From the earliest times on, a plethora of things have been either prohibited, or at any rate not customary in the waging of war. In their totality these prohibitions are known as *ius in bello*. Originally the *ius* in question was enshrined in custom, religion, or both. Later it joined all other kinds of law, becoming every bit as formal and elaborate. Here all we can do is to provide a very few examples of the subjects it deals with.

Some things were *never* allowed. The most important one is the sneak attack, treacherously conducted in violation of oaths, promises, and pacts. As when truces are broken; or prisoners, having been granted quarter, take up arms again; or fighters belonging to one party pretend to belong to the other. The fact that such things are prohibited does not mean that they are not done. What it does mean is that they have always been understood as contrary to the law of war, whatever its origin and whatever it might be called. Those who perpetrated them were punishable and, when the opportunity permitted, punished. Often this was done in the most barbarous and terrible ways anybody could think of.

Unfortunately such measures are sometimes hard to distinguish from ruses, stratagems, and surprise. In particular, commandos engaged on "special" missions have often worn enemy uniforms or, before there were uniforms, insignia to mislead the other side and take him by surprise. For example, some medieval scholars wondered whether Deuteronomy 22:5, which prohibits transvestism, also applied to men pretending to be women in order to launch a *coup de main*. These problems have always hampered attempts to deal with the matter in a just way. They are likely to continue doing so as long as wars are fought on earth.

The number of prohibitions that, reflecting historical circumstances, changed over time is so long that only a few can be mentioned here. Take the question of how sacred objects, buildings, and sites were supposed to be treated. The Greeks were tolerant in this respect. On pain of divine punishment, sometimes assisted by human punishment as well, temples, complete with the treasures they often contained, were supposed to be kept inviolate. The Romans went further still. They took their enemies' deities and brought them to Rome. The Anatolian goddess Cybele, the Egyptian one Isis, and the Persian god Mithra all had their worshippers there. Confucians and Buddhists might not go so far, but generally they were equally tolerant.

Monotheistic religions behaved differently. In the Old Testament any war between the Israelites and their enemies automatically involved their respective deities. Sacred objects of every kind were targets par excellence. The Prophet Samuel, hearing that the Holy Ark had been captured by the Philistines, fell off his chair, broke his neck, and died on the spot.[13] Each time the Israelites themselves won a victory the first thing they did was to exterminate their enemies' gods, root and branch. Christians and Muslims often did the same.

As late as the Thirty Years' War Protestants and Catholics regularly desecrated each other's places of worship. They might or might not take them over and re-consecrate them. The spread of secularization from 1650 on led to change. There was a growing tendency to leave places of worship alone—after all, if God did not intervene in human affairs, what was the point of desecrating or destroying or annexing sacred places and objects? Later art treasures, which in Greece and Rome had been seen as booty par excellence, also received immunity. In the West God, His buildings, and His servants still enjoy privileges. Not so in the Islamic world, where religion is often taken much more seriously. The way these things are handled forms an important, and growing, issue in the relations between the two civilizations.

Linked to this question was the treatment of enemy dead. Especially but by no means exclusively in tribal warfare, corpses were often mutilated. The objective was to inflict humiliation and/or prevent dead enemies from entering the afterworld. Perhaps the worst form of humiliation seems to have been cannibalism. By some accounts, defeated enemies were cooked, swallowed, digested, and excreted, though whether this ever happened in reality is moot. In ancient Greece the normal practice was to allow the defeated enemy to retrieve his dead and bury them. Much later, during the twentieth century, enemy bodies became the subject of several international treaties. They regulated questions such as registration, burial, notification of relatives, disposal of the dead soldiers' personal effects, and so on. Special organizations, coming under the auspices of the Red Cross, were established to administer them.

The question of who could be taken captive and how captives were to be treated gave rise to similar problems. As we saw, rarely did war end with every single leader, follower, man, woman, and child among the defeated being killed. Depending on the kind of society by which it was waged, and also on custom, different solutions were adopted. Leaders often formed first-rate targets—the Romans had a special medal for commanders who killed the enemy commander in combat. Those who were captured were put to death, often after being put on display, mutilated, and tortured.

Their normal fate was to be thrown off the Tarpeian Rock, an 80-foot-high cliff overlooking the Forum. But not all suffered this fate. One who escaped it was the Queen of Palmyra, Zenobia (r. AD 267–c.274). Zenobia rebelled against the Emperor Aurelian and was defeated. The early fourth-century historian Trebellius Pollio says she was saved, at least in part, by her "incredible sex appeal." Other leaders at other times were held in greater or lesser comfort until they paid ransom. At other times still their persons were considered more or less sacrosanct so that captured ones were set free, more or less. As happened first to Napoleon I on Elba and then to his nephew, Napoleon III, following his capture at Sedan in 1870.

Tribal societies almost always killed adult male captives, not seldom in gruesome ways. But they often allowed women, especially young and beautiful women, as well as children to live, treating them as booty and ultimately incorporating them. More sophisticated polities with stronger forms of government usually followed their example. However, instead of killing the men they might enslave them too. Both Greeks and Romans did so as a matter of course. During the Middle Ages the treatment of prisoners reflected their social status. Knights would be held for ransom and might be released on their word of honor to provide it. Commoners, who had neither honor nor gold, were more likely to be killed.

As late as 1618–48, armies still often took civilians prisoner. Rich men were captured to make them pay ransom. Rich women were captured with the same purpose in mind. In addition, poor—less often, it seems, rich—women could be captured for sexual purposes. A statuette by the German sculptor Leonhard Kern (1588–1662) shows a naked woman, her hands tied behind her back.[14] She is being led by a Swedish soldier pointing a dagger at her. All this was done on the orders, or at least with the connivance, of the commanders on the spot. Senior commanders also closed an eye, or else such things could not have happened as often as they did.

The person who did more than most to establish the distinction between combatants and noncombatants was, again, Hugo Grotius. Starting around 1660, the adoption of uniforms made the two groups much easier to tell apart. First bilateral treaties were drawn up with provisions for treating civilians as well as the wounded, the sick, and captives. Later they were systematized in the law of war as written down between 1868 and 1909 in particular (from the St. Petersburg Declaration, through the Hague Conventions of 1899–1907, to the Geneva Convention of 1909). The Geneva Conventions of 1949 then increased the protection afforded to these groups.[15] They also extended it to cover civil wars, including such as spill over into, or are subject to the intervention of, third parties. More changes followed the rise of feminism from about 1970 on. Previously women had enjoyed equal rights with men according to the groups to which they belonged. Now they were to be "especially protected" from sexual violence committed both by their captors and by their fellow prisoners. Pregnant women and mothers were also to enjoy certain privileges.

Space prohibits us from even trying to cover all the many other issues dealt with by *ius in bello* or, to use its modern name, international humanitarian law. Like *ius ad bellum*, it is governed by political imperatives, economic requirements, culture and custom, and the nature of the enemy. There are two reasons why it is critically important. First, had not *ius in bello* been observed to a greater or lesser extent, and in the absence of a clear definition of what victory means, every war would have to be fought to the bitter end. Doing so would have been expensive both in terms of casualties and of spoils lost. Second, he who violates the law and mistreats his enemies can expect to be punished by his own side, though that has always been rare, or, if he is captured, by the enemy.

Like any other kinds of law, *ius ad bellum* and *ius in bello* are some-times, some would say often, violated. The fact that, until recently, there was no impartial court to which victims could turn did not help. Even today, the court in question—the International Court of Justice in The Hague—is not very powerful. But that does not mean that, separately or together, law does not matter, even if it is very minimal and can safely be ignored both in theory and in practice. The law of war, Cicero wrote, is meant to preserve our humanity.[16] It seeks to prevent us from waging war in the manner of beasts, from allowing war to turn us into beasts. Of all the activities we humans engage in, war is the most likely to make us forget who and what we are. That is precisely why it needs to be, and usually is, subject to *some* kind of justice and *some* kind of law.

XI

Asymmetric War

1. The Principle of the Thing (b)

As noted in the introduction to this book, the term asymmetric war may mean two very different things. It may refer to war waged between polities each of which belongs to a different civilization. This happened, for example, when the Greeks fought the Persians, when first the Arabs and then the Mongols invaded Europe, when the Muslim Moguls took over Hindu and Buddhist India, and when Europe expanded into other continents. But the term "asymmetric war" may also mean a war between two different parties, one of which is so much weaker or stronger than the other that, by the normal rules of strategy, it should simply be no contest. Here the two problems will be treated in this order.

With Sun Tzu the idea that wars are waged within a single civilization is taken very much for granted. It goes a long way to explain the curious tit-for-tat quality in the conduct of operations so evident throughout his work. To be sure, some armies were better than others. Their commanders cultivated their humanity and justice and maintained their laws and institutions. Consequently they were better attuned to their troops, their people, and heaven itself. Being better attuned, they had a better understanding of the political and military situation. They were also better able to handle their spies, a difficult problem on which Sun Tzu puts great emphasis. Their maneuvers and stratagems were also more sophisticated, better adapted to local conditions and to the purpose at hand, and more successful.

In all this, as even some modern admirers have noticed, true creativity is hard to find. In a civilization such as the Chinese, one in which a personal God is lacking, history necessarily played a large role. For that reason, but also because technological change proceeded at a snail's pace, the ancients knew everything, war included, better. Every single one of the rules, stratagems, and tricks of the trade that fill the pages of *The Art of War* has existed from time immemorial. Given sufficient goodwill and effort, anybody could pick them up, study them, and master them. The more so because many scholar-soldiers, Sun Tzu himself included, had followers whom they taught and with whom they discussed these things. They were also itinerant, moving from one court to another in search of employment. Often they were acquainted with each other. As a result, we constantly find side A doing X. Side B responds by doing Y, and side A responds to the response by doing Z, and so on. Subsequent commentators, elaborating on the master's ideas, go even further in presenting war almost as if it were a game of *go* or chess. Only with real terrain, real obstacles (both natural and artificial), real weapons, real troops, and real bloodshed added.

Such methods were used by the various Chinese states to fight each other until, after some two and a half centuries, the one known as Qin prevailed and united the country. How commanders, using the same methods, fared against the "Barbarians" we do not know. Arguably, though, the failure of Sun Tzu and many of his followers to pay attention to the nature of the barbarians and comment on how waging war on them differed from other campaigns became a hallmark of much subsequent Chinese military thought. The barbarians were always poor and few in number. Materially and culturally, they could not compete with the mighty Chinese civilization-cum-state. Yet repeatedly they succeeded in overrunning the latter and conquering it. To this day, it is said, the inhabitants of "the Middle Kingdom," or "all under heaven" suffer from "Great Power Autism" (to use a term coined by the American strategist and pundit Edward Luttwak).[1] They are reluctant to come down from their pedestal and put themselves in other peoples' shoes. From Clausewitz down, many other writers on war have also failed to consider this kind of asymmetric war.

Let us now turn our attention to the second type of asymmetric war under consideration: war between two different parties in which one is so much weaker than the other that there should normally be no contest between them. In Clausewitz's favor it must be said that, though he only devotes a few pages to it, they are excellent indeed.[2] Yet the normal translation of the term he uses, *Volksbewaffnung*, as "the people in arms" is misleading. Its real meaning is "arming the people." As the name of the chapter before the one dealing with it, "Retreating into the Interior of the Country," implies, the model Clausewitz had in mind was Napoleon's 1812 campaign in Russia which, as a member of the so-called "German Legion," he witnessed. Later he lectured on what, at the time, was known as "little war." In both cases the government embraced its people and called it to help fight and drive out a foreign army that had invaded their country. In neither was there any question of insurgents rising *against* their own government which, in contrast to Clausewitz's examples, did everything in its power to disarm them.

In *The Art of War* the problem is more pronounced still. Here, the existence of terrorism, guerrilla insurgency, briefly what, today, we would call "sub-conventional" armed conflict, is not acknowledged. Everything is couched in terms of one state, one ruler, one commander, and one army confronting another of the same kind. Why that should be the case is hard to say, for surely the above-mentioned forms of war are as old as war itself. Long before the master's birth China must have had its share of them, as it has often done since. One explanation is that China, a Confucian society, has always seen itself as a single vast family whose structure and methods of government are mandated by heaven. Even more than in the West with its more individualistic approach to life, those who disturbed the cosmic order by taking up arms and rising against the emperor were considered criminals. On this view, it was not so much a question of fighting the rebels as of hunting them down like dogs.

As the term guerilla ("little war") implies, the reason why asymmetric wars by the weak against the strong and the strong against the weak were so often overlooked was because they were overshadowed by their bigger brothers. After all, the Latin term for war, *bellum*, is derived from *duellum*, a duel. When Clausewitz, on the first page of his book, said that war is nothing but a duel, he was simply repeating a phrase many of his predecessors had used. A duel, however, always implies a certain kind of equilibrium between those who fight it. When the imbalance is too great a fight is both unnecessary and impossible. But what happens when there is an elephant on one side and a gnat on the other?

As we said, this chapter deals with two different kinds of asymmetric war. The first—inter-civilizational war—is waged between polities belonging to different civilizations. The second is fought by the weak against the strong and, conversely, by the strong against the weak. The two asymmetrical forms can be combined in all sorts of ways. What gives them their importance is the fact that, from 1945 on, they have formed the vast majority of wars. Arguably their role in shaping global political life is much greater than that of the few remaining intra-civilizational "conventional" wars (i.e. wars fought *within* civilizations) that have long formed the backbone of military history and, perhaps even more so, of military-historical writing. It is likely to grow greater in the future. Those who ignore that fact, pretending that "it cannot happen here," are deluding themselves. Putting their heads in the sand, they risk being kicked in the behind.

2. War of Civilizations

How important the law of war really is becomes clear if we consider a war waged not within a single civilization but between two different ones. Doing so we find that war, like so many other things, is to a large extent governed not by any "objective" circumstances but by the minds of those who wage it. One might, indeed, argue that, consciously or not, the members of each civilization create their own "war convention." Some conventions are customary, others written. They define why it should be fought, what for, and so on; briefly, what counts as war and what does not.

It lies in the nature of things that belligerents belonging to the same civilization can usually find some common ground, tacit or explicit, on these questions. Not so those belonging to different ones. Not recognizing the same set of laws, they cannot stay within it even if they want to. Historically many peoples did not even subscribe to the modern idea of a common humanity that embraces us all, but saw themselves as the only real humans. With them, in other words, all wars waged against members of other communities or polities were inter-civilizational by definition.

The people of some civilizations required that war be opened with a more or less formal declaration, others did not. Some required that warriors wear uniform so as to identify themselves, others have never heard of such a thing. Another thing the war convention must do is to enable the two sides to communicate so as to ask and receive quarter. At present the white flag is universally used for the purpose. Formerly, though, people in each civilization had their own ideas on the matter. At the Battle of Cynoscephalae in 198 BC the Macedonian troops, tightly packed in a phalanx, were unable to defend themselves against the Roman legionaries with their short, deadly swords. At one point they tried to surrender in the only way they could, raising their pikes and holding them upright. Meeting Hellenistic troops for the first time, the Romans missed the sign.[3] They kept slaughtering their enemies until somebody explained it to them.

If Macedonians and Romans, living fairly close to each other, could misunderstand each other in this way, how much more so in cases when the distance, geographical and cultural, between the belligerents was greater. Consider the period of European imperialist expansion. For several centuries on end white men contacted people of other races all over the world. Often the outcome was total mutual incomprehension. Take the Spanish *conquistadores*' demand that the "Indians" in America convert to Christianity as a condition for not being fought and killed. What the Indians, seeing a black-dressed man standing in front of the ranks, waving a cross with an image of another man nailed to it, making strange gestures, and mumbling strange words may have thought is not recorded. They may well have concluded that the strangers were mad. The strangers, in turn, considered that Indians were not "person[s] capable of natural judgment sufficient to receive the faith, nor of the other virtues needed for conversion and salvation." Hostilities broke out, and the rest is history.

As Europeans continued to expand they became increasingly aware of the vast gaps that often separated their own laws of war from those of the Ottomans, Arabs, Indians, Chinese, Japanese, and Africans. That was one reason why, starting around 1750, they started codifying their own laws and calling them "international." Nevertheless, misunderstandings continue to abound. Arriving in Afghanistan in 2002, the Americans discovered that many adult males go about armed almost as a matter of course. Not only do they carry weapons, but they also fire them to celebrate the birth of a son, a wedding, and the like. How many "innocents" or "noncombatants," as they are known today, paid with their lives for this misunderstanding? And how often did insurgents, claiming to be noncombatants on their way to a party, turn their weapons against the occupation forces?

Even that only scratches the surface of the problem. Take intra-European war as it has developed from 1648 on. Waged by "powers" against one another, they tended to grow out of conflicts between rulers over relatively major political and economic issues. In this and other cases, many of those issues only affected the bulk of the population indirectly, at one or more removes. Often many, perhaps even most, of those who fought and died did not ask why and had only the foggiest idea of what the war was all about. "Freedom, or iron, or coal, or the devil knows what," as the Soviet writer Ilya Ehrenburg (1891–1967), referring to French and German soldiers in World War I, put it.[4]

Inter-civilizational wars are different. They may involve such issues; but they also may, and often do, witness clashes between different ways of life. In the words of the late Samuel P. Huntington: "People of different civilizations have different views on the relations between God and man, the individual and the group, the citizen and the state, parents and children, husband and wife, as well as differing views of the relative importance of rights and responsibilities, liberty and authority, equality and hierarchy. These differences are the products of centuries. They will not soon disappear. They are far more fundamental than differences among political ideologies and political regimes."[5]

The most sensitive problems are those involving sex and gender. The people of some civilizations see nothing wrong with homosexuality or pederasty. Others abhor them. Some, having practiced female circumcision for millennia, consider it a sacred duty. Others think of it as a monstrosity. Or note the contrast between the West and the Islamic world. In the former, women have long enjoyed relative freedom to go about and do as they please. In the latter they have often been, or chosen to be, secluded. They wore veils, stayed at home, and either were prevented from meeting or preferred not to meet strangers. The early decades of the Arab–Israeli conflict (c.1900–48) illustrate these differences very well. The Arabs disliked the independence of Jewish women who, they claimed, were running about half-naked. They feared that their own women might follow suit. Many of the Jews, for their part, were afraid of what they felt the Arabs, if given the opportunity, might do to the women of their community.

Close mutual acquaintance can alleviate such differences. But it can also make them worse. Nor is "globalization" reducing their importance. To the contrary: massive migration often introduces them into "advanced" countries where, it was thought, they had long been overcome. In one sense, these and similar issues are less "serious" than those secular reasons for which European and Western rulers have long gone to war, at least in the era since the Peace of Westphalia (1648) ended Europe's religious wars. Rarely do these issues confront nations with the choice between life and death (unless it is spiritual death). On the other hand, they concern far more people; and often do so in a much more intensive, personal way.

Such different attitudes to the most basic things in life can turn into a tinderbox which the smallest spark will set alight. Once war breaks out each side will find it hard to classify the enemy, understand what he is trying to do, and fight him. The most elementary ideas as to what is and is not right, is and is not permissible, will evaporate. That, again, is why inter-civilizational wars are so often fought *à outrance*. Where understanding is lacking there will be a tendency to see the other side as brutes whom only brute force can bring to heel. If we treat an enemy as a fanatic he will become a fanatic. If we hate and despise him, then as surely as night follows day he will hate and despise us right back.

Different civilizations will lead to different objectives, strategies, and different notions as to what constitutes victory. On the other hand, since war forces both sides to study each other and imitate each other, it may also cause those differences to diminish over time. But not before plenty of blood has been shed. Nor have nuclear weapons caused these problems to lose their importance. It is not self-evident, as some believe, that nuclear weapons are simply nuclear weapons, deterrence is deterrence, compellence is compellence, and so on.[6] Indeed it would be a miracle if leaders, raised and educated on the knees of different civilizations, did *not* see these things differently. The "signals," deliberate and accidental, verbal and non-verbal, which leaders from one civilization send their colleagues in other civilizations are often prone to being misunderstood. Indeed, in war games designed to explore the issue, they almost always have been.

Fortunately there are three factors at work which, though they do not eliminate the danger, seem to reduce it. First, building a nuclear arsenal is a slow process, often made slower by the vast expense involved. Publicly available information indicates that only the U.S., driven by total war, did it in less than five years. The Soviet Union, employing truly draconian methods, took some seven. All the rest, being much smaller and often less advanced, took about a decade, sometimes more. Even when they really put their mind to it (as was the case in Pakistan, whose then Prime Minister Zulfikar Ali Bhutto, promised that his people would "eat grass" to build the bomb).[7] And even when they received outside help to develop a nuclear weapon, as most did. Which means that those in charge have always had plenty of time to think things through. And do so, moreover, with their predecessors' experience in front of their eyes.

Second, as Thomas Schelling notes, in the game of nuclear deterrence and compellence, as in strategy as a whole, time will create precisely the kind of tit-for-tat situation so prominent in *The Art of War*. The longer the engagement, the more alike the opponents will become. Football and basketball will merge into foot-basketball. As the ruckus around Iran in 2012–15 showed once again, the most dangerous time is when a country is about to shift from non-nuclear to nuclear status. That done, as time passes, mutual understanding—"calculability," as it is sometimes known—should improve. As far as post-1945 history allows us to judge, it *has* improved, at least to the extent that no nuclear war has broken out.

Third, and most important, is sheer fear. Fear of a holocaust so terrible that the living, both "victors" and losers, will envy the dead. The first thing anybody who seeks to "use" nuclear weapons for deterrence or compellence needs is fearlessness, real or fake. To repeat, however, since Hiroshima fear of these weapons seems to have increased, not decreased. Just how the leaders of North Korea and Iran, let alone those of Al Qaeda and similar organizations, see the matter we do not know. But we *do* know that, in the past, nuclear weapons made even two of the most absolute, bloodthirsty, and paranoid dictators of all, Stalin and Mao, think twice. That was true both before they had them and afterwards.[8]

However this may be, the jinnee is now out of the bottle. Either we learn to live with the jinnee, or it will kill us all.

3. Weak against Strong, Strong against Weak

Jomini, at the beginning of his most important book, *Précis de l'art de la guerre* (1836), says that wars are made by governments in the name of their polities. In fact, as we saw, that was not always true even in his own time. Let alone throughout history. Many tribal peoples did not have a government, at any rate not in the modern sense of the term. In many other cases organizations that did not form polities, or commonwealths, or states fought either each other or the polities in question.

The people who made up these organizations were usually called by all kinds of bad names, among which "rebels" is one of the more complimentary. But that does not change the fact that they *were* organizations and that they used violence not just for private gain but as a means for achieving political ends. Like other warriors, they felt justified, if not by the law of the governments under whom they lived, then by some higher cause, or law, or truth. For that cause, or law, or truth, they were prepared to risk their lives, often more readily than their opposite numbers in the "forces of order." Often they also enjoyed significant popular support.

By and large such wars follow the principles of strategy, though on a smaller scale and with some modifications. Those who waged them being much weaker than their opponents, their first rule has always been to avoid fighting in the open. Seeking to prolong the struggle, they hid in difficult terrain such as mountains, swamps, forests, and less often, deserts. Doing so implied a fairly large country and, in many cases, open borders through which the insurgents/guerrillas/terrorists could receive aid. They could also conceal themselves by dispersing among the people. In fact their ability to obtain the people's support and disappear into them when necessary was their single most important asset.

From beginning to end, the first objective of guerrillas and their ilk has always been to gain time. When the enemy advanced, they retreated. When the enemy retreated, they advanced. Breaking cover, they focused on the enemy's weaker links, provoking him, surprising him, raiding him, harrying him, and hopefully driving him crazy as a horse is driven crazy by flies. Next they disappeared. If, using such methods, they made the enemy strike around blindly, killing civilians and noncombatants and driving the population into the arms of the insurgents, so much the better. All this, to the accompaniment of intense propaganda, psychological warfare, and political measures designed to attract the population and control it.

Sometimes the rebels, or whatever they were called, succeeded in their purpose. Perhaps more often they were suppressed, allowing the modern territorial state to emerge from about 1500 on. However, at some time after World War I, for reasons that are not altogether clear, the balance started to tilt. First, as happened in Morocco during the Rif Rebellion of 1921–6, well-established colonial powers such as France and Spain found that holding down their subjects had become much harder. It now required hundreds of thousands of troops and several years of sustained warfare including the use, on the Spanish side, of poison gas. Even Hitler's Germans, by no means a soft-hearted lot, were unable to put down various resistance movements in many of the countries they had occupied.

Then the dams broke. For fear of nuclear escalation, interstate war went into decline. Its place was taken by intrastate warfare. In such warfare nuclear weapons, owing to their very power and the durability of their effects, were largely irrelevant. The first to feel the impact were the imperialist powers. Seeking to hold on to their colonies, they waged numerous campaigns against their rebellious subjects. So savage were some of the wars as to approach genocide. To no avail, as one after the other the imperialists were forced to leave.

The Americans in Vietnam and the Soviets in Afghanistan fared no better. Quite a few of the operations other than war, or peacekeeping operations, or whatever they may have been called, launched from 1990 on did not succeed either. Nor were "advanced" Western countries the only ones to fail. The Chinese Nationalist Government failed to put down the Communists (1927–49). The same applied to the Indonesians in East Timor (1975–99) and the Vietnamese in Cambodia (1978–90).

Among the best indicators of the new importance of asymmetric war are legal changes. For most of history sharp distinctions were drawn between warriors and rebels. The former were often supposed to have at least some rights, the latter none. Now various international agreements, especially Article 1 (4) of the 1977 Additional Protocol 1 to the Geneva Conventions, were signed to "include armed conflicts in which peoples are fighting against colonial domination and alien occupation and against racist regimes in the exercise of their right of self-determination." The outcome was to remove, at least to some extent, the difference between those fighting for the state and those fighting against it.

To be sure, each state claimed that, whatever else, those who rose against it were nothing but criminals. That was how the British in Kenya treated the Mau Mau and the Israelis the Palestinians, to mention but two cases out of many dozens. This made the application of the law extremely uneven. On the whole, though, its role has been growing. On one hand, commanders and troops are constantly in danger of being photographed by one of the many small gadgets on the market. On the other, never before have they been expected to document their every move as they are now. During the American war in Afghanistan, each time an Afghan was killed the responsible officer had to fill in a form at least five pages long, detailing everything from the dead person's posture and dress to the ambient temperature.[9] Behind him armies of lawyers, word processors at the ready, were waiting to pounce on every word and every letter.

No two of these wars were identical. Still they did have some things in common. Most were waged by states and their regular armed forces against enemies who, certainly at the outset, could only be described as miserable bands or gangs. Not having a code of military law, those bands were unable to enforce formal discipline. They relied almost solely on their members' devotion combined, when necessary, with rough, ready, and not seldom extremely severe justice. Precisely because they did not have formal judiciary powers, they had to set an example when the occasion demanded it. They were poor in numbers, experience, training, weapons, and money. At one point during the 1964–79 Rhodesian War, so poor was the Zimbabwe

African National Union that it could not pay the telephone bill of its delegation in London.

Yet the "bands" and "gangs" in question did enjoy some advantages. Being small, their financial and logistic requirements were limited. They were agile and could hide among the civilian populations. The most important advantage they enjoyed was their status as underdogs. It gave them no choice but to use every available means. That in turn gave them a certain freedom that their stronger opponents, who did have a choice, did not have. They were, in other words, often better able to influence hearts and minds, both in their own society and among bystanders. Even when they broke the laws of war. A law in whose formulation they had no part, and which had been enacted by states specifically to help combat groups like themselves, among other things.

They could, for example, take hostages or ignore the distinction between combatants and noncombatants. These and other advantages help explain why, from the Jews in Palestine who rid themselves of the British Mandate in 1944–8 through to the Taliban in Afghanistan, most of these movements attained their objectives, more or less. And they did so in the teeth of everything their much more powerful opponents could do.

For the latter, excuses were not lacking. Many defeated commanders cited the support that the insurgents, or rebels, or guerrillas, or terrorists, received from across the border. Some blamed the politicians who failed to provide direction, others the media that undermined public trust. Often they spoke of "lack of coordination," thus putting the blame on somebody or something else. The chain of command, designed for conventional warfare, was too tall and too cumbersome. The character of the struggle, which flared up now here, now there, encouraged senior officers to engage in micromanagement and get in each other's way, paralyzing their subordinates. Intelligence and operations were not sufficiently integrated, leading to long reaction times. Or defeat was put at the door of the "strategic corporal." In an environment that often demanded not just military skill but political, social, cultural, and psychological finesse, he was out of his depth. Either he overreacted or he did not react at all.

Many of these factors attended many of these wars. A few may have afflicted them all. Yet the most important one is usually ignored. For the weak, even failure, provided it is publicized, is success. It proves he is still there. For the strong, even success is failure. If his troops kill their enemies they will be accused of brutality that is both criminal and counterproductive. If they allow themselves to be killed in turn, they will be accused of incompetence. Either way, the outcome will be an investigation and at least the threat of punishment. Fear of punishment will make everybody lie to everybody else, all the time. As the war in Vietnam illustrated very well, gradually everybody—enlisted personnel, officers, senior commanders, the media, public opinion, and politicians—will become demoralized.

Given time, a strong army fighting a weak one will *become* weak, and the other way around. To this extent war is the continuation, not of policy/politics but of sport. What to do? Here we shall tackle the question by focusing on two armies in two campaigns. Each in its own way succeeded in avoiding the problem and emerged victorious as a result. The first campaign is the one President Hafez Assad waged in Syria in 1982. At the time, his regime was meeting growing opposition that did not make its future appear rosy. Part of the opposition to his regime had to do with the fact that he and his principal collaborators were members of an unloved minority sect, the Alawites. Part had to do with Islamic opposition to the secular character of the state. To make things worse, since 1976 many of Assad's forces had been involved in the civil war in Lebanon. In early 1982 that country was also threatened by an Israeli invasion.

The Muslim Brotherhood, a religious organization with branches in every Arab country, mounted a well-organized and effective terrorist campaign. As Assad's regime disintegrated and his own life was threatened, he resorted to desperate measures. The center of the rebellion was the city of Hama. In the midst of a ferocious repression campaign 12,000 soldiers, commanded by Assad's brother

Rifat, surrounded Hama. They started combing the city house by house, making arrests. As they did so, some 500 *mujahidin*, or holy warriors, launched a counterattack, reportedly killing some 250 civil servants, policemen, and the like.

The uprising gave Rifat and Hafez the excuse they had been waiting for. Relying mainly on their most powerful weapon, heavy artillery, their troops surrounding Hama opened fire. Anywhere between 10,000 and 30,000 people, many of them women and children, were killed. What followed was even more important than the killing itself. Far from apologizing, Rifat, when asked how many people he and his men had killed, deliberately exaggerated their number. As his reward, he and his principal subordinates were promoted. Survivors told horrifying tales of buildings that had collapsed on their inhabitants and trenches filled with corpses. For years thereafter, people passing where the city's great mosque had been looked away and shuddered.

Previously Syria had gone through one coup after another. Now it enjoyed thirty years of peace. Five cardinal principles contributed to victory. First, the blow was prepared in secret. When it came, it hit like a thunderbolt. Second, ere it began a bait was used to lure the enemy out of his hiding places. Third, it was very powerful—better too powerful than too weak. Fourth, it was short. Fifth and most important, it was done in public and without apology. The most powerful blow in the world will not do away with everybody on the other side. Much if not most of the impact is psychological. An apology, as by saying one is sorry for the non-combatants who died, will weaken that impact, perhaps fatally so.

The second campaign is the one in Northern Ireland. The "troubles" in Ireland go back all the way to King Henry II (r. 1154–89), the first English monarch to try and take over the island. In January 1969 they broke out again. As bombs demolished parts of the infrastructure such as electricity pylons and water pumps, and as opposing demonstrators fought street battles with each other, they quickly escalated.

From this point the situation went from bad to worse. In a single night's "battle" (Belfast, August 14–15, 1969) four policemen and ten civilians were killed and 145 civilians were injured. Property damage was also extensive. From January to August 1971 alone 311 bombings caused over one hundred injuries. In 1972 the number of bombings increased to well over 1,000. The IRA also extended its operations from Northern Ireland into the United Kingdom proper. A peak was reached on "Bloody Sunday," January 30, 1972. On that day the army's attempt to deal with street fighting in Londonderry left thirteen people dead.

Had things been allowed to continue in the same way, no doubt the British attempt to hold on to Northern Ireland would have ended as so many others had, i.e. in complete defeat. If this did not happen and the outcome did not follow the usual pattern, then perhaps there are some things to be learnt from the effort. Here we shall focus on the things the British Army, having learnt from their own experience and that of others, did *not* do.

First, never again did they fire into marching or rioting crowds. However violent the riots and demonstrations, they preferred to employ less violent means, causing far fewer casualties. Second, not once did they use heavy weapons such as tanks, armored personnel carriers, artillery, or aircraft to repulse attacks and inflict retaliation. Third, never did they inflict collective punishments, such as imposing curfews, blowing up houses, and the like. Instead, they positioned themselves as the protectors of the population, not its tormentors; by doing so, they prevented the uprising from spreading. Fourth and most important, by and large the army stayed within the law.

From time to time, this rule was infringed upon. Even without breaking the law, interrogation techniques could be intimidating enough. There were clear violations of civil liberties. Torture and false accusations were used to elicit information and obtain convictions. A few known IRA leaders, identified and tracked in foreign countries, were shot, execution-style, in what later became known as "targeted killings." On the whole, though, the British played by the rules, refusing to be provoked. Even after terrorists blew up the seventy-nine-year-old Earl of Mountbatten, the Queen's uncle, in his yacht. Even after they planted a bomb that demolished part of a hotel where Mrs. Thatcher was due to speak. And even after they fired mortar rounds at a cabinet meeting at 10 Downing Street.

The real secret behind the British success was iron self-control. As Machiavelli noted, no other use of power is so impressive to those who are exposed to it. Rooted in a certain kind of society, the manifestations of self-control are patience, professionalism, and discipline. A look at the butcher's bill will confirm this claim. In most counterinsurgencies, for every member of the "forces of order" who lose their life, at least ten insurgents are killed. Taking "collateral damage," as the phrase goes, into account, the difference is much larger still. By contrast, the struggle in Northern Ireland left 3,000 people dead. Of those about 1,700 were civilians, almost all of whom fell victim to terrorist bombs. Of the remaining 1,300, 1,000 were British soldiers and just 300 terrorists. As one British officer told me, "That is why we are still there."

The first method maximizes the use of force whereas the second minimizes it. To that extent they are opposites. Yet they also have something in common; to wit, the fact that, in both cases, the regulars focused on the importance of time as a vital factor, so as to prevent it from doing its demoralizing work. The Syrian Army did so by ending the campaign almost before it began. The British one achieved the same effect by making the forces immune to its passing—"courageous restraint," as the approach has been called. But what happens, as is often the case, when people do not have what it takes to follow either course consistently? To speak with a 2004 *Washington Post* headline referring to U.S. counterinsurgency operations in Iraq, they switch "from killing to kindness" and back again.[10] Doing so, they demoralize their own troops, which are left without guidance, and encourage the enemy.

Rarely do military measures alone suffice to decide an asymmetric struggle of this kind. The irregulars, or guerrillas, or terrorists, or whatever they are called, must draw the population to their side, or else they cannot win. The counterinsurgents, or whatever they are called, want to control it so as to deprive the insurgents of the "sea" in which, to speak with Mao, they "swim." On both sides this requires excellent intelligence about the "sea" in question. This is something the irregulars, thanks to their roots in the population, often find easier to obtain than their opponents, especially if the latter are foreigners. Next they use propaganda and intimidation to draw those populations to their side. Both try to deprive each other of their allies, bribe them, divide them, and, if possible, turn their factions against each other. Since irregulars of all kinds seldom form a monolithic block, they are especially vulnerable to such techniques.

A campaign not decided by any of these methods may peter out, especially if the insurgents fail to enlist public opinion on their side and/or run out of resources. Or else it may escalate, emerge into the open, and turn into a full-scale civil war—what Mao called "the third stage"—subject to the normal principles of strategy. Often, though, the above-mentioned processes make time work for the guerrillas and against their enemies. They enable the former to prevail even before that stage is reached. To quote the American statesman Henry Kissinger (1923–), guerrillas, as long as they do not lose, win. Their opponents, as long as they do not win, lose.[11]

The final point is that, for insurgents and counterinsurgents alike, in such a conflict politics becomes so all-pervasive that it engulfs the military conflict, becoming indistinguishable from it even at the lowest levels. In this way asymmetric war is the mirror image of total war in which, by contrast, war becomes so all-pervasive that it completely engulfs the political side of things, thereby becoming indistinguishable from it. Both forms of war reveal Clausewitz's famous dictum for what it really is. Not a universal truth, but a special case; one that stands between two extremes.

Perspectives

Change, Continuity, and the Future

1. Change

Historically speaking, the changes war has undergone are nothing short of momentous. Yet military change cannot and does not proceed on its own. It is rooted in a whole host of political, economic, social, and cultural factors and affects those factors in return. That is why predicting it is enormously complex, often all but impossible.

Some of the most important changes have taken place in the field of organization. We do not know who the first warriors were. Yet we do know, or think we know, that they were organized in fairly loose tribal bands. With the "agricultural revolution" and the more centralized, more hierarchical polities it helped create came militias, feudal levies (whose members fought in return for the usufruct of land), mercenaries, and standing armies. Often different kinds of units were combined in one way or another and fought side by side, a method that did not contribute to their cohesion or their reliability.

The French Revolution reintroduced the principle, often used by ancient city-states and their medieval successors but long abandoned, of general conscription. Fifty years later the rise of the railways enabled governments and states to supplement conscription by putting in place systems for mobilizing reservists. These systems remained in use during both World Wars and beyond. At peak, they enabled some of their owners to create forces numbering almost 10 percent of their populations and keep them in the field for years on end. Starting around 1970, things changed again. Cultural, social, technological, and economic changes caused the role of conscripts and reservists to go down, whereas that of professional armies went up.

The development of these various forms of organization has not always done away with simpler, less centralized, and less hierarchical ones. Often the forms in question were adopted, if that is the right word, by those who could not afford to set up large, permanent, and expensive armed forces. From around 1500 on, it was the armed forces of the emerging modern states that held the upper hand, so that by 1914–45 a mere handful of them shared practically the entire globe between them. The only forces that could take them on and stop them were others of the same kind. Whether that will continue to be the case in the future remains to be seen.

Next, economic development. The simplest tribal societies were able to wage war at almost no economic cost. This gave them an enormous advantage over more settled ones and helps explains why, as late as AD 1650, some could still play an important role in international politics. Generally, though, economic and military power went hand in hand. Other things being equal, the richer the society the more soldiers it could afford to train, maintain, and equip.

The onset of the industrial revolution caused the gap between rich and poor societies, "civilized" and "barbarian"—as they were once called—to grow enormously. At one point, so great was this gap that the former, using a small fraction of their resources, were able to take over the latter almost at the drop of a hat. Since then the gap has narrowed, but it has not closed. "Developed" countries are still able to threaten most "developing" ones with blockade and invasion. The opposite is not true. But is there an upper limit beyond which riches, instead of adding to a polity's military power, detract from it? The past suggests there is; as to the future, time will tell.

Third, technological change. Over the ages, the growth in the power, speed, range, accuracy, and so on of weapons, weapon systems, and other kinds of equipment has been spectacular. Starting with the invention of the telegraph and reaching to the most recent sensors, data links, and computers, the ability to gather information, transmit it, and process it has grown much faster still. All these capabilities keep increasing with no limit in sight. Some weapons have become so powerful that, for fear of blowing up the world, they cannot be used.

Around 1914, technological change led to a revolution in logistics. Previously the most important logistic requirements were food and fodder. Next, as the battlefield became mechanized, the requirement for other kinds of supply skyrocketed. Many newly needed items were highly specialized. Gathering them from the countryside was impossible. This in turn made bases, lines of communication, and transport more important than ever. Conversely, one reason why the operations of all sorts of terrorists, guerrillas, and insurgents have often been successful is because, as the proverbial "handful of rice" indicates, their logistic requirements are minuscule by comparison.

Repeatedly in the past, technology has enabled war to expand into additional environments. From the land to the sea; from the surface of the sea to the depths; into the air; into space; and into cyberspace. But for technology war at sea, under the sea, in the air, and in outer space would not be possible. Cyberspace would not even exist. Will future wars extend into additional environments and dimensions? There are precedents: Albert Einstein in 1905–15 added a fourth dimension to the existing three. There may be more waiting out there. A subterranean one of the kind described in more than one nineteenth-century novel, perhaps? Or an ethereal one through which thought miraculously passes from one person to another?

That is speculation. Yet one thing seems clear. Should new environments and dimensions be discovered or constructed, it will be only a matter of time, and not a very long time either, before technology helps fill it with "ghastly dew."[1] That was what the English poet Lord Tennyson (1809–92), anticipating the conquest of the air, saw raining down from "the central blue." Indeed there is a sense in which man's conquest of any new environment remains incomplete until it is fought in, and over, too.

Originally the tools used for war were the same as those used for hunting. Later the two separated. As the role of hunting as a source of food declined, military technology drew ahead. During antiquity non-organic sources of energy such as windmills and waterwheels were invented. They were, however, fixed in geographical space and could not be used in the field. As a result, military technology started lagging behind the best civilian technology. It was only the advent of the industrial revolution, especially the internal combustion engine, which caused things to change again.

As research and development turned into sustained, well-organized processes from about 1890 on, military technology, benefiting from large investments, drew ahead of its civilian counterpart. Not by accident did the Wright brothers turn to the military as their first presumptive clients. The invention of the microchip around 1980 again changed the equation. Devices produced by civilian companies working for the civilian market were often better than those many militaries around the world possessed, including some of the most modern ones. This revolution has just begun. Its full impact will probably be revealed over the decades to come.

As military technology changed it was accompanied by changes in doctrine, training methods, and so on. Often they also caused armed forces to lay their plans in different ways, deploy in different ways, operate in different ways, and fight in different ways. So much so, in fact, that terminology had to change as well. As happened around 1780 when the term strategy, long forgotten, was dug up and, having lost its Greek moorings, started to be used to describe the higher operations of war.

With strategy came a host of other terms—bases, objectives, lines of communication, interior and exterior lines, to mention but a few. Later, grand strategy was added. Some also inserted another layer, operations, between strategy and tactics. Change also caused the meaning of other terms, such as ground battle, to change. For centuries on end it referred to massive clashes between the two sides' main forces. Now that those forces have become so dispersed that they seldom meet, most of the time it refers to mere skirmishes.

In 1945 the first nuclear weapons were detonated. Nothing that had happened in military history, perhaps in history *tout court*, before that date was nearly as important. Nothing that has happened, or is likely to happen, after it is nearly as important either. For the first time, humanity created a situation where it could destroy itself by a "wargasm," as it has been called. Since then fear of escalation has not only ended major war between major powers, but this effect has spread outward from them like ripples in a pond. Such war has been replaced by deterrence on one hand and compellence on the other. Insecurity has given birth to a certain kind of security. Yet there is no guarantee that nuclear weapons will never be used. They may be. Should that happen, we shall *really* have entered a different world—assuming there will still be a world at all.

Changes in the law of war have been numerous and momentous. There was a time when many, perhaps most, men on the defeated side were executed. That done, the best women and children could hope for was to be kept alive and enslaved. Other societies, instead of putting the men to death, enslaved them. Nobody did so more systematically, and on a larger scale, than the ancient Greeks and Romans. At times the law of war reflected class divisions. At others class mattered comparatively little. The years after 1700 gave rise to the idea that the wounded form a separate category with rights of their own. Later, even the dead acquired some rights.

Some, including Clausewitz, tended to play down the importance of the law of war. Others, such as Cicero, emphasized it. Never mind that, by modern standards, many aspects of Roman warfare were barbaric. In part it is a matter of perspective. It is no accident that Cicero was primarily a lawyer, Clausewitz a soldier. Respect for the law is notoriously hard to measure. To the extent that it can be measured, though, there is some reason to think that recent developments, primarily the ubiquity of cameras and the ease by which information is recorded and disseminated, has caused the role of law to grow. An important sign pointing in this direction was the establishment, in 2002, of the International Criminal Court in The Hague. For the first time in history, it provides a permanent forum where war criminals can be and are being tried. In many quarters, concern as to what it may do has been rising.

As conventional war wanes, the number and importance of wars waged by the strong against the weak and the weak against the strong appears to be growing. Along with other forms of irregular warfare, in time they may supplant the conflicts of the past as well as the armed forces that used to fight them. The outcome will be new kinds of organization. Hezbollah and Al Qaeda, both of which incorporate religious, military, economic, charitable, and, not least, criminal aspects, and both of which are at home with modern technology, may be among the prototypes. Before we address the future of war, though, we must first take a look at the things in war that have *not* changed.

2. Continuity

Military or civilian, organizational or technological, change does not proceed evenly at all times and places. Sometimes it moves, if it does at all, at a glacial pace. At other times, with war stepping on the pedal, it greatly accelerates. Some polities remained wedded to their existing forms of war for millennia on end. Others changed them, often repeatedly, in surprising directions and at surprising speed. Much the most important driving force behind the process is the "swing effect" previously mentioned. In the life-and-death struggle that is war, the choices are stark. Either change so as to keep up with your enemy and overtake him if possible. Or else, *vae victis*.

Nevertheless, a great many things have not changed, nor do they seem about to change. First come the causes of war, about which more in the next section. Then there are its most essential characteristics. Among them are its nature as a two-sided struggle which is, or should be, subject to the rules of strategy; its violence, always tending towards escalation and making it hard to control; its role as an instrument of policy, however defined, without which it is a senseless thing, without an object; the fact that, unlike crime, it is considered acceptable and even laudable by large, if not all, segments of the society that wages it and in many cases the enemy too; and the fact that it is not an individual activity but a collective one.

The challenge war presents and the demands it makes on those who wage it do not change either. The most important challenge is that of the responsibility that rests on those at the top. Next come the uncertainty, the friction, and the *Strapazen*, including the greatest challenge of all: fear of a quick, brutal death. Since war first began, none of these has changed one bit. As long as war remains, none will. The same is true of the qualities and processes needed to cope with these *Strapazen* and overcome them, such as courage, determination, cohesion, organization, training, discipline, and leadership. In sum, fighting power.

To be sure, in every modern armed force the percentage of the troops directly exposed to enemy fire has been declining from the middle of the nineteenth century on. Many now "participate" in "combat" from thousands of miles away. Some do so by programing computers, others by launching missiles or operating drones. They do so while seated in air-conditioned rooms, some of which are located so deep underground that they can only be demolished by a direct nuclear hit. They watch their consoles and manipulate their controls. All this, without the slightest danger to themselves.

These facts raise some disturbing questions. Is the activity in which drone operators in particular are engaged really war? Or is it merely high-tech butchery? If it is the latter, how can it be justified, both morally and in law? Is war going to lose its human element? The answers to the first two questions are moot. As to the third, it seems to be negative. As experience shows, it is not impossible for the side that has and uses drones and their relatives, robots, to be fought to a standstill by those who do not. Just look at what the U.S. has (not) succeeded in doing to Islamic State since that organization first started making its mark back in 2006.

One of the principal goals societies set themselves when going to war has always been to gain resources. Be it in the form of access to hunting and grazing grounds or water; or women and children; or slaves; or agricultural land; or mineral resources; or hoards of gold and silver; or a host of other things. Such aspirations are still alive and well, especially but not exclusively in intrastate warfare. As of the early years of the twenty-first century there is much talk of "resource" and "climate" wars. Some believe that water shortages will lead to armed conflict. Others think more in terms of energy or certain rare, but critically important, raw materials. The kind of resource varies and will no doubt continue to vary. However, economic factors remain as important as ever.

Starting about 1830, military technology has been advancing at a furious, perhaps accelerating pace. Used against those who do not have it, or who have an inferior version of it, technology is enormously powerful and gives its owners an enormous advantage. So large is it that, for a time, it appeared irresistible. As the English writer Hilaire Belloc put it, referring to colonial warfare around 1900 when the gap between Europe and the rest peaked: "when everything is said and done we've got/the Maxim gun, and they have not."[2] Used against an enemy with similar technology at his disposal, technology's impact has always been much more limited. One reason why it has been limited is the fact that other factors have retained their importance, as they still do.

Moreover, as Clausewitz says, the fundamental principles of strategy are dictated less by the tools it uses than by its own nature. That explains why technology has failed to have much effect on them; also, why not one sentence Sun Tzu wrote about strategy twenty-five centuries ago has become dated. This is shown, among other things, by the fact that, broadly speaking, the principles he set forth are applicable to war at sea, an environment he did not address; and even to war in the air, space, and cyberspace, which he could not and did not envisage.

And what about nuclear weapons? Superficially they appear to have changed everything. So much so that certain science-fiction writers considered banning them essential if they were to continue writing about war at all. But there are limits to what they have done. Huge as their shadow is, underneath a great many things go on as usual. As the English pioneer of armored warfare and military commentator J. F. C. Fuller said in 1946, one does not abolish war by threatening to eradicate cities. Much less by eradicating them, which will merely invite retaliation. War is a flexible and inventive beast. Like some mythological shape-shifter, it will adapt itself without giving up its essential nature.

The law of war has undergone any number of important changes. However, the need for it has not. Now as ever, war involves a radical and often sudden switch from a situation in which many things are prohibited to one in which they are accepted, commanded, and commended. Now as ever, it is necessary to define who has the right to wage it and who does not, on whose orders, against what opponents, for what reasons, with what objectives, by what means and methods, and so on. Briefly, to define what does and does not count as war. Those who stay within the limits deserve to be decorated; those who violate them, to be punished (as, sometimes, they are). Countless practical questions arise and need to be resolved, from the rights of prisoners of all ages and both sexes all the way to the rights of terrorists.

Just how important the law really is, and how great its impact, remains as moot as it has ever been. The truth is that, after two centuries and two millennia respectively, both Clausewitz and Cicero still have right on their side. An army which, for one reason or another, has its hands tied behind its back by law (and lawyers) may end by losing even to opponents weaker than themselves. That is because the latter, operating on the principle that necessity knows no law, feel free, are free, to do what has to be done and do it. On the other hand, a polity which allows its troops to do anything without regard to law will turn into a herd of beasts. In time, it will likely cease to be a polity at all.

To repeat, nuclear proliferation and the decline of large-scale conventional war seem to cause the number and importance of the first kind of asymmetric war, i.e. that waged by the strong against the weak and the weak against the strong, to grow. As to the second kind, i.e. intra-civilizational war, it has always existed. Such wars center on the most personal issues of all: be they religious, or social, or cultural, or psychological. Evolutionary pressures such as sex and propagation may also be involved. That is why they have always been, and remain, particularly destructive and particularly barbaric.

3. Does War Have a Future?

"Does War Have a Future?" was the title of an essay published in *Foreign Affairs* in October 1973. That very month the Yom Kippur Arab–Israeli War, the largest and most modern in two decades, broke out. Less than a year after that a coup overthrew the dictatorial government of Portugal. Previously Lisbon had been trying to put down uprisings in its colonies, Angola and Mozambique. The outcome was civil war in both countries, which lasted for decades and cost hundreds of thousands of lives.

Nothing daunted, subsequent authors kept hammering on the theme. A few used the exact same title. Nor was the 1973 essay by any means the first. As early as the 1580s Jean Bodin, in *An Easy Method for the Study of History*, made many of the arguments bandied about today. Those who, hoping to put an end to war, thought they could see that end coming, included the economists Friedrich List (1789–1846) and Norman Angell (1872–1967, who received a Nobel Peace Prize for his efforts); the philosophers John Stuart Mill (1806–73) and Herbert Spencer (1820–1903); the anthropologist Margaret Mead (1901–78, who, in 1940, argued that warfare was "only an invention")[3] . . . Each time their hopes were dashed.

Early in the twenty-first century, the statistics show, the average person is less likely to die in war than in any previous period. That is wonderful news, except that it may be due not to our becoming more peaceful but simply to a population explosion without precedent in history. Furthermore, war has always been a periodic activity punctuated by more or less prolonged, more or less complete, pauses. It has been claimed that, since c.1400 BC, less than 10 percent of all years were entirely peaceful. Thus any respite we have been granted— who knows why?—may be temporary. Look at the stock exchange. As each bubble in turn fills with hot air, the gurus say that *this* time things are different. Only to have the next crash prove they are not.

As far back as the 1890s, the philosopher Herbert Spencer compared war with slavery—both were based on coercion rather than free exchange. Hence, he claimed, it would soon follow the latter into the dustbin of history. But slavery has *not* disappeared. Early in the twenty-first century there are some thirty million slaves in the world. In some cases war itself is re-establishing slavery as boys and girls are abducted. The former to act as fighters or workers; the latter as servants, concubines, or prostitutes. Furthermore, slavery, like war, has a long history. Having gone through a vast variety of forms, it is very hard to define. In many countries "guest workers" have their passports taken away and live under conditions not so very different from slavery. They can even be bought and sold, in a few cases legally so. The term "wage-slaves" speaks for itself. If we took these and many other forms of forced labor into account, our figure for the number of slaves worldwide would be much larger.

Furthermore, war is always a deliberate act, the outgrowth of policy/politics. The line between war and politics may be fairly clear, as Clausewitz's dictum, referring to conventional interstate war, implies. Or it may be almost nonexistent, as (for different reasons) happens both in total war and in insurgency/counterinsurgency. Either way, the one way to abolish war is by making sure it can no longer serve the objectives of policy/politics. The two principal objectives of war have always been to assuage fear and satisfy greed. Fear and greed in turn are stimulated by shifts in relative power—military, political, and economic. History is a cakewalk; over time it makes fear or temptation, sometimes both, increase or decrease. As in Hobbes' time, there is no supreme court capable of adjudicating disputes in an "impartial" manner. Let alone an international police force sufficiently powerful to enforce its decisions on any but the weakest "sovereign" polities of all.

From Immanuel Kant and Tom Paine in the late eighteenth century, all the way to some modern political scientists, much has been written about the reluctance of democracies to fight each other. That may be true, more or less. But democracies have certainly not been loath to fight against, or in, non-democratic countries. At times they claimed that doing so was their duty. Back in the 1990s, it looked as if most countries were on their way to becoming democracies and history had ended. But the ten years' grace that the collapse of the Soviet Union gave the world were soon over. Power politics, assuming they had ever ceased to operate, came roaring right back. To be sure, fear of a nuclear holocaust still prevented major war between major powers, even such as are governed by dictators. But that fear could not and did not prevent many smaller wars in many other places in the world.

Historically, many wars were launched with an eye to gaining wealth. The list of rulers and polities which, whether or not that had been their original intention, profited from war starts with the ancient Middle Eastern monarchies and ends with the U.S. in both World Wars. Had it not been for President George Bush, Sr.'s decision to intervene, Saddam Hussein's 1990 invasion of Kuwait would have proved highly profitable. Should any Southeast Asian ruler decide to conquer the tiny, more or less defenseless Sultanate of Brunei, surely the enterprise will be much more profitable still.

Even assuming most interstate wars are no longer economically profitable, they may still bring other benefits. As Friedrich Nietzsche wrote, victory is the best cure for the soul.[4] It can and will increase the winners' deterrence power, making them less vulnerable to attack. As, for example, happened to Britain following its 1982 triumph over Argentina. In theory, modern rulers are not supposed to personally profit from the wars they wage on their country's behalf. But whether or not they intend to profit, in practice they often do exactly that. If not in financial terms, then by gaining enhanced prestige and eligibility for office. Without the timely invasion of the Falklands and her own response to it, Margaret Thatcher might not have been re-elected. Even in Hanoi during the late 1970s, following a quarter-century of ferocious warfare, there were rich men with fat bellies.

But who says wars have to be waged between states? Since at least 1945, the vast majority, including many of the most bloody, have been waged *inside* them. At times governments fight non-state organizations and the other way around. At others times, all sorts of non-state organizations fight each other. Starting from nothing, the heads of the various groups and organizations quite often become filthy rich. Yasser Arafat (1929–2004), founder and longtime leader of the Palestinian Liberation Organization, is said to have died a multibillionaire. So, apparently, did Angolan guerrilla leader Jonas Savimbi (1934–2002). Many of their followers, receiving payment and/or preying on the surrounding society, also find war a good way to make a living. Some may be better off than their opposite numbers who, serving the state, must survive on the crumbs the latter throws their way. That is one reason why such wars often last as long as they do— those who wage them have every reason to protract them as much as they can.

We now come to the thorniest questions of all: where may the better angels of our nature be found? Have they taken over, are they taking over? Can any changes occurring within us, if they do, bring war to an end? The idea that we are becoming, or capable of becoming, better, kinder, gentler, less rapacious, and less cruel is a product of the late Enlightenment. "All men will become brothers," wrote Friedrich Schiller in the *Ode to Joy* in 1785. Four years later the French Revolution broke out, drowning Europe in rivers of blood.

Nor did the slaughter end in 1815. Still to come were the 1854–60 Taiping Rebellion in China (20–30 million dead), the 1861–5 American Civil War, the 1864–70 war between Paraguay, Argentina, and Uruguay (relative to the number of combatants the most deadly one in modern history), and the Austro-Prussian and Franco-Prussian Wars. Some late nineteenth- and early twentieth-century colonial wars, notably the Belgian one in the Congo and the German one in

Namibia, wiped out entire populations. Next came the Russo-Japanese War, the two Balkan Wars, World War I, the Sino-Japanese War (including "The Rape of Nanking"), the Italian war in Ethiopia, the Spanish Civil War, and World War II. In the period 1945–2013 alone there have been some 200 wars. Among the most important ones were the Chinese Civil War, the Korean War, Biafra (in Nigeria), Algeria (two separate wars), the War in Vietnam, Angola, Mozambique, Sri Lanka, Ruanda, the Sudan, and Zaire. Not to mention smaller wars and massacres in Afghanistan, Chad, Lebanon, Iraq (twice), Sierra Leone, Somalia, and Yugoslavia.

Looking back over the last two centuries, the one thing more common than predictions about the end of war has been war itself. And this list does not even include mass murders such as Stalin's Great Terror, the Holocaust, the Great Leap Forward, the Cultural Revolution, and Cambodia's killing fields. Time after time the trumpet has sounded. Time after time masses of people, most of them ordinary, well-behaved people no different from anyone else, have responded and shed their veneer of civilization. The hounds of hell have repeatedly been unleashed and unspeakable crimes have repeatedly been committed, sometimes with an underlying rational purpose, but sometimes driven by nothing more than sheer sadism. The number of those who, during the twentieth century, died by "politicide," as it has been called, is estimated at a quarter billion. Whoever, after all that, still believes in moral progress must be brave indeed.

Throughout history, the largest and deadliest wars have always been the ones that the great powers waged against each other. If, relative to the global population, the number of those who died in war has in fact declined, then that is mainly because major war between such powers has become all but extinct. That in turn is due, not to angelic intervention but to raw fear lest war, following its natural course, may escalate. Possibly ending in a holocaust without parallel in history; and possibly bringing history to a spectacular end.

However, as the above list shows, there are many forms of war that nuclear weapons can *not* prevent. In the so-called "developing" world, hardly a day passes without some new armed conflict, large or small, breaking out. Many of the wars in question were launched by leaders seeking this or that objective. As they dragged on, though, the distinction between means and ends tended to be lost. Foreign actors, including not just states but corporations seeking to protect their investment or make a profit, became involved. So did mercenaries. Often the outcome was an enormously complex, monstrously bloody and destructive, ever-changing, yet more-or-less permanent morass, sucking in leaders, followers, and noncombatants, and offering them no escape. Actors became victims and victims, actors. Both looted the civilian population, killing those who tried to resist. As in the Thirty Years' War, war and crime became fused. If anything, the importance of such conflicts is growing. Zaire (1987–), Sierra Leone (1991–2003), Somalia (1991–), and Afghanistan (1981–8, 2002–) are good cases in point. So, as of 2016, are Libya, Syria, and Iraq.

In most of the "developed" world things are different. There governments use Tweedledum and Tweedledee, their social services and their security organs, to hold their populations in check. To the point where most people, never having heard a shot fired in anger, see war as a spectator sport. What it really means they can't imagine. Confronted with it, often they stop their eyes and ears. But is there any country so "advanced," so rich, so homogeneous, and so wallowing in its content, as to be immune to war? How about climate change, if such a thing exists? How about economic crises, past, present, and future? How about minority groups who, right or wrong, feel discriminated against? How about friction between different cultures, often involving the things people hold nearest and dearest? How about countless laws which, in the name of decency, democracy, diversity, equality, equity, the environment, health, humanity, morality, safety, and of course political correctness, are shackling us hand and foot? Has Lao Tzu's dictum that the larger the number of laws, the larger also the number of those who break them ceased to apply?[5] And how about religious differences which, experience shows, are quite capable of unifying these grievances and others like them, weaving them into a crown of thorns, and thrusting it onto society's head?

To quote Francis Bacon once again, there never has been, nor will there ever be, a shortage of "seditions and troubles."[6] A world court capable of adjudicating them is nowhere in sight; hence, surely, some cases will continue to be settled, if at all, by force of arms. Does that mean war will continue to afflict humanity forever? Not necessarily. To be sure, politics—international anarchy on the one hand and structural flaws inside communities on the other—play a large role. Still, at the most basic level, war feeds on our drives and emotions. Chief among them are hatred, aggression, rage, revenge, and what Nietzsche calls the "will to power." All heartily reciprocated, all rebounding on ourselves.

Should those drives and those emotions be erased from our souls, then rulers and polities will find that going to war has become impossible. That is one reason why armies of physicians, psychiatrists, psycho-pharmacologists, brain scientists, and geneticists are trying to modify, canalize, rein in, and perhaps eliminate them. Not, they claim, without some success. Day by day, in many countries, countless fetuses are aborted because of defects, both physical and mental, discovered during pregnancy. Will the day come when they are killed because, say, tests show that they carry a "pugnacious" gene?

Nor are fetuses the only group affected. Many mentally disturbed people as well as some criminals routinely undergo electrical, chemical, and hormonal treatments. More treatments, based on nanotechnology, are coming. Some are voluntary, others not. Children, who depending on which country we are talking about, form between one quarter and one half of the entire population, are close behind. The goal is to fashion their emotions and moods so as to suit society; round pegs into round holes. Who comes next?

But does war reflect only the worst parts of our nature? The attacker, wrote Lenin in his notes on Clausewitz, always wants peace.[7] He wants to occupy our country, take away our freedom and our property, and, should we try to resist, kill us too. That is why he so often claims to be coming in peace. Hitler wanted peace, except that he wanted Danzig and Poland and Scandinavia and the Low Countries and France and the Balkans and Russia first. So did many others before and after him.

"When they poured across the border/I was cautioned to surrender/ this I couldn't do" sang the Canadian singer-songwriter Leonard Cohen (1934–). Is defending home and hearth such a bad thing? Should we always turn the other cheek? Some Christian, Hindu, and Buddhist sects preferred being killed to killing. So did Jesus, Saint Francis, and Mahatma Gandhi. Religious movements could do so because they felt protected by the larger societies surrounding them. As to the three leaders, suppose they had been responsible for some kind of organized polity. In that case their behavior, far from being heroic, could have, almost certainly would have, been considered criminal, if not treasonous.

War is evil. But is it exclusively and entirely evil? Aren't some things even worse? How about injustice, how about persecution? To avoid war, should Abraham Lincoln have permitted slavery to stand? Should Britain have accepted Hitler's offer and made peace in 1940? Have freedom and dignity ceased to matter? Have survival and comfort become the only goals of life? How far shall we allow ourselves to be regulated in their name? Shall we allow our bodies and our minds to be re-engineered? Doesn't war engage our enterprise, our courage, our desire to test ourselves to the utmost? How about love, out of which wells our willingness to suffer and to sacrifice, and—should there be no other way—to die for something or somebody else? What to do with these, our best qualities? Store them? If so, will they be there when we need them? Not having cultivated them, shan't we be easily defeated? Finally, since the evil and the good in us are inter-twined, won't the methods used to suppress the former do away with the latter as well?

Already our every step and every word can be monitored, not only without our consent but without our knowledge too. Shall we also have our thoughts and feelings controlled? Is being made into marionettes a price worth paying for the peaceful life? Where would humanity be without war? In some hybrid of *Brave New World* and *1984*, perhaps? These are questions each society and each person should decide for themselves. As long, that is, as the prevailing climate of opinion still allows them to be asked at all.

As for us, *our* trip through fearful war is closed and done.

NOTES

Introduction

1. "War is a matter of vital importance to the state." Sun Tzu, *The Art of War*, S. B. Griffith, trans. Oxford, Oxford University Press, 1963, p. 63.
2. Clausewitz says that the law in question...C. von Clausewitz, *On War*, M. Howard and P. Paret, eds., Princeton, NJ, Princeton University Press, 1976, p. 76.
3. "The greatest name of my military empire." Quoted in N. D. Wells, "In the Support of Amorality."
4. Had experience been enough. Karl Demeter, *The German Officer-Corps in Society and State, 1650–1945*, New York, Praeger, 1965, p. 67.
5. "We do not have a steady supply of Hindenburgs." J. V. Stalin to L. Z. Mekhlis, 1942, quoted in https://en.wikipedia.org/wiki/Lev_Mekhlis.
6. From knowledge to capability is a great step. Clausewitz, *On War*, p. 147.
7. "Peruse again and again." *Napoleon's Military Maxims*, W. E. Cairnes, ed., Mineola, NY, Dover, 2004, p. 80.

Chapter I

1. "The most hateful god." *Iliad* 5.583.
2. "Doomed to make war." *Correspondence Letters between Frederic II and M. de Voltaire*, Th. Holcroft, ed., London, G. J. G. and J. Robinson, 1809, p. 7.
3. "There is defeat and cunning." Quoted in Ling Yuan, *The Wisdom of Confucius*, New York, Random House, 1943, pp. 157–8.
4. "Ceases only in death." Th. Hobbes, *Leviathan*, London, J. Palmenatz, ed., Collins, 1962, p. 135.
5. "Lift up the ensign from afar." *Isaiah* 5:26.
6. "Slept with one of the Trojan men's women." *Iliad* 2.355.
7. Genghis Khan. See on this D. L. Hartl, *Essential Genetics: A Genomic Perspective*, Boston, Jones, and Bartlett, 2011, pp. 159–60. http://news.nationalgeographic.com/news/2003/02/0214_030214_genghis.html.
8. "His hands dripping with gore." *Iliad* 11.169.
9. "Dry-mouthed, fear-purging ecstasy." E. Hemingway, *For Whom the Bell Tolls*, London, Arrow, 2004, p. 243.

10. J. Glenn Gray, *The Warriors*, New York, Harper & Row, 1970, p. 56.

11. "Great fun" and following quotes: J. M. Wilson, *Siegfried Sassoon, The Making of a War Poet*, London, Duckworth, 1998, pp. 179–80, 221, 268, 291, 317, 319, 510.

12. When Stalin said, "dance." W. Taubman, *Kruschchev: The Man and His Era*, New York, Norton, 2003, p. 211.

13. Sun Tzu on estimates: Sun Tzu, *The Art of War*, pp. 6–7.

Chapter III

1. Fighting is to war what cash payment is to commerce. Clausewitz, *On War*, p. 148.

2. Dismissed it as "childish." R. Hofmann, *German Army War Games*, Carlisle Barracks, PA, Army War College, 1983, pp. 29–30.

3. The fourth-century BC ... Livy, *Roman History*, viii.1.

4. As Clausewitz says ... Clausewitz, *On War*, pp. 87–8.

5. Absolute war: Clausewitz, *On War*, p. 78.

6. "The Cradle of Bolshevism." Quoted in R.-D. Mueller and G. R. Ueberschaer, *Hitler's War in the East*, New York, Berghahn, 2002, p. 104.

7. "The God of War" Clausewitz, *On War*, p. 583.

8. A similar story ... D. Pietrusza, *The Rise of Hitler & FDR*, Washington D.C., Rowman & Littlefield, 2016, p. 150.

9. Rock soup method. G. S. Patton, *War as I Knew It*, New York, Fontana, 1979, p. 120.

10. Waging war is like trying to walk in water. Clausewitz, *On War*, p. 583.

11. *Strapazen* (physical effort) of war. Clausewitz, *On War*, p. 115.

12. Josephus Flavius, *The Jewish War*, Book 3, Chapter 5.

13. Will turn into a childish game. Plato, *Laws* 796 and 830c–831a.

14. Iron-fisted training. R. Hoess, *The Commandant*, New York, Duckworth, 2012, locs. 266–71.

Chapter IV

1. Each Mamluk could take on three French soldiers. Quoted in B. Colson, ed., *Napoleon on the Art of War*, Oxford University Press, 2015, p. 81.

2. "The Rise of the Staff in the Western Way of War," MilitaryHistoryOnline. com, at http://en.citizendium.org/wiki/Napoleonic_military_staff.

3. Complained that his contemporaries. M. de Saxe, *Reveries on the Art of War*, Mineola, NY, Dover, 2007, pp. 36–8.

4. "Love and be loved." Plato, *Republic*, 460B.

5. "A battalion of 200 ..." G. H. von Berenhorst, *Betrachtungen ueber die Kriegskunst*, Leipzig, Fleischer, 1797, vol. 2, p. 424–5.

6. "On a hot summer day ..." *Wenn die Soldaten* ... H. Dollinger, ed., Munich, Brueckmann, 1974, p. 61. The translation is mine.

7. Th. Campbell, ed., *Frederick the Great, His Court and Times*, London, Colburn, 1848, vol. p. 138.
8. *"Pourtant, ils ne peuvent pas voler."* Quoted in J. B. Vachée, *Napoléon en Campagne*, Paris, Legaran, 1900, p. 195.
9. "In peace there's nothing so becomes a man..." *Henry V*, III.1.

Chapter V

1. "A free creative activity." Heeresdienstvorschrift 300, *Truppenfuehrung*, Berlin, Mittler, 1936, p. 1. The translation is mine.
2. J. Clement, *The Lieutenant Don't Know*, Philadelphia, PA, Casemate, 2014, p. 249.
3. The greater the distance, the greater the cost. Sun Tzu, *The Art of War*, p. 72.
4. As late as the mid-eighteenth century... B. Davies, *Empire and Military Revolution in Eastern Europe*, London, Continuum, 2011, p. 69.
5. $53 billion on foreign intelligence alone. *CIA Handbook*, Washington D.C., 2014, p. 249.
6. Neither divine omens, nor crystal balls... Sun Tzu, *The Art of War*, p. 145.
7. Without anger and without favoritism. Tacitus, *Annales* 3.9.

Chapter VI

1. "System of expedients." H. von Moltke, "Ueber Strategie" (1871), in *Militaerische Werken*, Berlin, Mittler, 1891, vol. 2, part 2, p. 293.
2. At the highest level. Quoted in Colson, ed., *Napoleon on the Art of War*, p. 83.
3. The highest triumph is won. Sun Tzu, *The Art of War*, p. 77.
4. Frederick Lanchester argued that... F. W. Lanchester, "Mathematics in Warfare," J. R. Newman, ed., *The World of Mathematics*, New York, Simon and Schuster, 1956, vol. 4 pp. 2138–57.
5. Unlike other generals. Napoleon on War, at http://www.napoleonguide.com/maxim_war.htm.
6. The course Moltke, Sr. in particular recommended. Moltke, *Miltitaerische Werke*, vol. 3, part 2, p. 163.
7. He immediately knew all was lost. J. Jackson, *The Fall of France*, Oxford, Oxford University Press, 2003, p. 9.
8. The first rule of strategy... Clausewitz, *On War*, p. 204.
9. Like a lion or like a fox. N. Machiavelli, *The Prince*, Harmondsworth, Penguin, 1969, p. 99.
10. A good general might wage war... De Saxe, *Reveries*, p. 121.
11. Liddell Hart developed this idea... B. H. Liddell Hart, *Strategy*, London, Faber and Faber, 1954, *passim*.
12. A really great victory... A. von Schlieffen, "Cannae Studien," *Gesammelte Schriften*, Berlin, Mittler, 1913, vol. 1, p. 262.

13. It is for breaking the rules...D. MacArthur, *Reminiscences*, Annapolis, MD, Naval Institute Press, 1964, p. 264.
14. She is fickle...Machiavelli, *The Prince*, p. 133.

Chapter VII

1. The idea of invading the U.S....A. Hitler, speech of 28.4.1939, at http://comicism.tripod.com/390428.html.
2. In war between two powers...Francis Bacon, "Essays Civil and Moral," in *The Works of Francis Bacon*, B. Montague, ed., London, Care & Hart, 1844, vol. 1 p. 39.

Chapter VIII

1. "The young aviator lay dying." The full song is available at http://www.theaerodrome.com/forum/showthread.php?t=2774.
2. H. G. Wells' novel. H. G. Wells, *The War in the Air*, London, Bell & Sons, 1907.
3. "Space Pearl Harbor." J. M. Stouling, "Rumsfeld Committee Warns against 'Space Pearl Harbor,'" *SpaceDaily*, 11.1.2001, at http://www.spacedaily.com/news/bmdo-01b.html.
4. "Information War." See e.g. R. C. Molander, A. Riddle, and P. A. Wilson, *Strategic Information Warfare*, Santa Monica, CA, RAND, 1996.
5. Stuxnet. See on this K. Zetter, "An Unprecedented Look at Stuxnet, the World's First Digital Weapon," *Wired*, 11.3.2014, at http://www.wired.com/2014/11/countdown-to-zero-day-stuxnet/.
6. Clausewitz's dictum that surprise...Clausewitz, *On War*, pp. 198–9.
7. Snowden revelations. See C. Steele, "The 10 Most Disturbing Snowden Revelations," *PC News*, 11.2.32014, at http://www.pcmag.com/article2/0,2817,2453128,00.asp.

Chapter IX

1. "Another such victory..." Plutarch, *Life of Pyrrhus* 21.8.
2. "A *Canticle for Leibowitz*." Lippincott & Co., Philadelphia, PA, 1960.
3. "Star Wars" speech. The speech, which was delivered on 23.3.1983, is available at http://www.atomicarchive.com/Docs/Missile/Starwars.shtml.
4. Is about 15,000. Federation of American Scientists, *Status of World Nuclear Weapons 2015*, at http://fas.org/issues/nuclear-weapons/status-world-nuclear-forces/.
5. Welcome Dr. Strangelove. See, for the various strategies described in this chapter, L Freedman, *The Evolution of Nuclear Strategy*, New York, St. Martin's, 1984.
6. Unanimously recommended...R. M. Gates, *Duty: Memoirs of a Secretary at War*, Kindle Edition, 2014, loc. 10752.

7. Permissive Action Links. See, on them, Anon. "Principles of Nuclear Weapon Security and Safety," 1997, at http://nuclearweaponarchive.org/Usa/Weapons/Pal.html.
8. The Kargil War. See V. P. Malik, *The Kargil War*, New Delhi, IDSA, 1999.

Chapter X

1. "Those who excel in war." Sun Tzu, *The Art of War*, p. 88.
2. Clausewitz on his part did devote...Clausewitz, *On War*, p. 76.
3. "Sufficiently known." Hobbes, *Leviathan*, p. 143.
4. "And utterly destroy..." 1 *Samuel* 15:3.
5. The Murngin people of Arnhem Land. W. Lloyd Warner, *A Black Civilization*, New York, Harper, 1937, pp. 174–7.
6. The most extreme form of war...*Numbers*, chapters 31 and 32.
7. Subsequent Jewish scholars...See on this A. Grossman, "Maimondies and the Commandment to Conquer the Land," 17.9.2013, at www.etzion.org.il/vbm/archive/11.../29milchama.rtf.
8. Only then did he feel...*Iliad* 24.26.
9. Bodin in his famous...J. Bodin, *On Sovereignty*, J. H. Franklin, ed., Cambridge, Cambridge University Press, 1992, pp. 1–126.
10. "The first servant of the state." Frederick II, *Réfutation de Machiavel*, in *Oeuvres*, Berlin, Decker, 1857, vol. 8, pp. 169 and 298.
11. Virgil in the *Aeneid*. Book VI.
12. Drawing on previous Roman law...see, for St. Augustine and St. Thomas Aquinas, F. H. Russell, *The Just War in the Middle Ages*, Cambridge, Cambridge University Press, 1975, pp. 16–39, 258–91.
13. The Prophet Samuel...1 *Samuel* 25.
14. A statuette by the...An image of the statuette is available at https://www.google.co.il/search?hl=en&site=imghp&tbm=isch&source=hp&biw=1097&bih=559&q=Leonhard+Kern+&oq=Leonhard+Kern+&gs_l=img.12.0i3ol2.2137.2137.0.4484.1.1.0.0.0.0.129.129.0j1.1.0....0...1ac.2.64.img.0.1.129.pClFHpCcoMQ#imgrc=CcVZAb8TuOFWNM%3A.
15. The Geneva Conventions of 1949. The texts are available at https://www.icrc.org/en/war-and-law/treaties-customary-law/geneva-conventions.
16. The law of war...Cicero, *De officiis* 1.11.

Chapter XI

1. "Great Power Autism." E. N. Luttwak, *The Rise of China vs. the Logic of Strategy*, Cambridge, MA, Belknap, 2012, *passim*.
2. A few pages...Clausewitz, *On War*, pp. 479–83.
3. Meeting Hellenistic troops...Polybius, *The Histories* 18.26.10–12.
4. "Freedom, or iron, or coal." I. Ehrenburg, *Thirteen Pipes* (1923), Pipe No. 4.

5. "People of different civilizations..." S. P. Huntington, "The Clash of Civilizations," *Foreign Affairs*, 72, 3. summer 1993, p. 25.
6. It is not self-evident...See L. Sondhaus, *Strategic Culture and Ways of War*, London, Routledge, 2006.
7. "Eat grass." The exact time and place when the statement was made is disputed. See "Storing up trouble: Pakistan's nuclear bombs," *The Guardian*, 3.2.2011, at https://www.theguardian.com/commentisfree/2011/feb/03/paki stan-nuclear-bombs-editorial.
8. Think twice. See on this J. L. Gaddis, ed., *Cold War Statesmen Confront the Bomb*, Oxford, Oxford University Press, 1999, pp. 39–61, 194–215.
9. During the American war in Afghanistan...Lt. Jeff Clement, personal communication, p. 153.
10. "From killing to kindness." *Washington Post*, 4.12.2004.
11. "Guerrillas, as long as they..." H. A. Kissinger, "The Vietnam Negotiations," *Foreign Affairs*, 47, 2, January 1969, p. 2.

Perspectives

1. "Ghastly dew." Lord Alfred Tennyson, "In Locksley's Hall" (1842), available at http://www.poetryfoundation.org/poem/174629.
2. "We've got the Maxim gun..." H. Belloc, *The Modern Traveler*, London, Arnold, 1898, p. 42.
3. "Warfare is only an invention..." M. Mead, "Warfare is Only an Invention—Not a Biological Necessity," *Asia*, 1940, pp. 402–5.
4. "Victory is the best cure for the soul." *Daybreak*, Cambridge, Cambridge University Press, 1982, fifth book, aphorism No. 571.
5. The larger the number of laws...Lao Tzu on government, available at http://www.sacred-texts.com/tao/salt/salto8.htm.
6. "Seditions and troubles." F. Bacon, "Of Seditions and Troubles," in *Francis Bacon: The Major Works*, B. Vickers, ed., Oxford, Oxford University Press, 1996, pp. 366–71.
7. The attacker...*Lenin's Notebook on Clausewitz*, D. F. Davis and W. S. C. Kohn, eds., Normal, IL, Illinois State University, n.d., at http://www.clausewitz.com/bibl/DavisKohn-LeninsNotebookOnClausewitz.pdf, p. 167.

INDEX